Studies in Computational Intelligence

Volume 637

Series editor

Janusz Kacprzyk, Polish Academy of Sciences, Warsaw, Poland
e-mail: kacprzyk@ibspan.waw.pl

About this Series

The series "Studies in Computational Intelligence" (SCI) publishes new developments and advances in the various areas of computational intelligence—quickly and with a high quality. The intent is to cover the theory, applications, and design methods of computational intelligence, as embedded in the fields of engineering, computer science, physics and life sciences, as well as the methodologies behind them. The series contains monographs, lecture notes and edited volumes in computational intelligence spanning the areas of neural networks, connectionist systems, genetic algorithms, evolutionary computation, artificial intelligence, cellular automata, self-organizing systems, soft computing, fuzzy systems, and hybrid intelligent systems. Of particular value to both the contributors and the readership are the short publication timeframe and the worldwide distribution, which enable both wide and rapid dissemination of research output.

More information about this series at http://www.springer.com/series/7092

Xin-She Yang
Editor

Nature-Inspired Computation in Engineering

 Springer

Editor
Xin-She Yang
School of Science and Technology
Middlesex University
London
UK

ISSN 1860-949X   ISSN 1860-9503   (electronic)
Studies in Computational Intelligence
ISBN 978-3-319-80757-7   ISBN 978-3-319-30235-5   (eBook)
DOI 10.1007/978-3-319-30235-5

This Springer imprint is published by Springer Nature
The registered company is Springer International Publishing AG Switzerland

Preface

Nature-inspired computation has become increasingly popular in engineering and is thus leading to a paradigm shift in problem-solving and design approaches. Many design problems in engineering are complex with multiple nonlinear constraints, and their design objectives can often be conflicting under stringent design requirements. Such problems can be very challenging to solve because of their complexity, nonlinearity and potentially high-dimensionality. Traditional algorithms, especially gradient-based algorithms, can struggle to cope and the results are not satisfactory. New and alternative methods are highly needed.

In the last two decades, nature-inspired computation and optimization algorithms have demonstrated their effectiveness as an alternative set of design tools to deal with such complex design problems. As a result, the popularity of nature-inspired optimization algorithms has started to increase significantly in recent years. These nature-inspired algorithms include particle swarm optimization, differential evolution, cuckoo search, firefly algorithm, bat algorithm, bee algorithms and ant colony optimization as well as others. The key features of these algorithms are their simplicity and flexibility, which enable them to highly adaptable and sufficiently efficient to deal with a wide range of optimization problems. In practice, they are also easy to be implemented in almost any programming languages, and all these factors have contributed to their popularity in engineering.

A majority of nature-inspired optimization algorithms can be said to belong to the class of swarm intelligence (SI) based algorithms. Swarm intelligence concerns the high-level behaviour arising from simple interactions among multiple agents. Though the main characteristics of swarming behaviour may be drawn from different sources of inspiration in nature, the procedures and steps of each algorithm can be very different. However, they seem to use the self-organized abilities of complex systems based on simple interaction rules. Since most algorithms treat problems under consideration as a black box, it is possible for such algorithms to deal with complex problems with different properties whether continuous, discrete or mixed. In a broad sense, swarm intelligence is part of evolutionary computation paradigm and bio-inspired computation is part of nature-inspired computation;

however, there is still some confusion about some terminologies in the literature. For example, there are some overlaps and there are no agreed standard definitions about bio-inspired computation, nature-inspired computation, metaheuristic and computational intelligence. Therefore, we will not enter the debate about what the right terminology or subject fields should be, but we will rather focus our attention on nature-inspired computation and its applications in engineering.

The diversity and rapid advances in nature-inspired computation have resulted in a much richer literature. Therefore, a timely review is necessary to summarize the latest developments in terms of algorithm developments and their applications in engineering. Algorithms and topics include discrete firefly algorithm, discrete cuckoo search, plant propagation algorithm, parameter-free bat algorithm, gravitational search, biogeography-based algorithm, differential evolution, particle swarm optimization and others. State-of-the-art applications and case studies include vehicle routing, swarming robots, discrete and combinatorial optimization, clustering of wireless sensor networks, cell formation, economic load dispatch, metamodeling, surrogate-assisted cooperative co-evolution, data fitting and reverse engineering as well as other real-world applications.

As a timely review volume, this book can be an ideal reference for researchers, lecturers, graduates and engineers who are interested in latest developments in nature-inspired computation, artificial intelligence and computational intelligence. It can also serve as a reference for relevant courses in computer science, artificial intelligence, machine learning, natural computation, engineering optimization and data mining.

I thank our editors, Drs. Thomas Ditzinger and Holger Schaepe, and staff at Springer for their help and professionalism. Last but not least, I thank my family for the help and support.

London   Xin-She Yang
December 2015

Contents

Contributors

Ali R. Al-Roomi Dalhousie University, Electrical and Computer Engineering, Halifax, NS, Canada

Roberto Carballedo Deusto Institute of Technology (DeustoTech), University of Deusto, Bilbao, Spain

Fernando Diaz Deusto Institute of Technology (DeustoTech), University of Deusto, Bilbao, Spain

Mohamed E. El-Hawary Dalhousie University, Electrical and Computer Engineering, Halifax, NS, Canada

Bouazza Elbenani Department of Computer Science, Mohammed V University, Rabat, Morocco

Eid Emary Faculty of Computers and Information, Cairo University, Giza, Egypt

Iztok Fister Faculty of Electrical Engineering and Computer Science, University of Maribor, Maribor, Slovenia

Iztok Fister Jr. Faculty of Electrical Engineering and Computer Science, University of Maribor, Maribor, Slovenia

Akemi Gálvez Department of Applied Mathematics and Computational Sciences, University of Cantabria, Santander, Spain

Xingshi He College of Science, Xi'an Polytechnic University, Xi'an, People's Republic of China

Andrés Iglesias Department of Applied Mathematics and Computational Sciences, University of Cantabria, Santander, Spain; Department of Information Science, Faculty of Sciences, Toho University, Funabashi, Japan

Alan Kwan School of Engineering, Cardiff University, Cardiff, UK

Haijiang Li School of Engineering, Cardiff University, Cardiff, UK

Yang Liu School of Engineering, Cardiff University, Cardiff, UK

Salvatore Marano Department of Computer Engineering, Modeling, Electronics, and Systems Science, University of Calabria, Rende, CS, Italy

Uroš Mlakar Faculty of Electrical Engineering and Computer Science, University of Maribor, Maribor, Slovenia

Eneko Osaba Deusto Institute of Technology (DeustoTech), University of Deusto, Bilbao, Spain

Aziz Ouaarab LRIT, Associated Unit to the CNRST (URAC 29), Mohammed V-Agdal University, Rabat, Morocco

Nunzia Palmieri Department of Computer Engineering, Modeling, Electronics, and Systems Science, University of Calabria, Rende, CS, Italy; School of Science and Technology, Middlesex University, London, UK

Yacine Rezgui School of Engineering, Cardiff University, Cardiff, UK

Abdellah Salhi Department of Mathematical Sciences, University of Essex, Colchester, UK

Birsen İ. Selamoğlu University of Essex, Colchester, UK

Marwa Sharawi Faculty of Computer Studies, Arab Open University, Cairo, Egypt

Muhammad Sulaiman Department of Mathematics, Abdul Wali Khan University Mardan, Mardan, Khyber Pakhtunkhwa, Pakistan

Giuseppe A. Trunfio DADU, University of Sassari, Alghero, Italy

Xin-She Yang School of Science and Technology, Middlesex University, London, UK

Manal Zettam Department of Computer Science, Mohammed V University, Rabat, Morocco

Nature-Inspired Optimization Algorithms in Engineering: Overview and Applications

Xin-She Yang and Xingshi He

Abstract Nature-inspired computation has become popular in engineering applications and nature-inspired algorithms tend to be simple and flexible and yet sufficiently efficient to deal with highly nonlinear optimization problems. In this chapter, we first review the brief history of nature-inspired optimization algorithms, followed by the introduction of a few recent algorithms based on swarm intelligence. Then, we analyze the key characteristics of optimization algorithms and discuss the choice of algorithms. Finally, some case studies in engineering are briefly presented.

1 Introduction

Optimization is everywhere and important in many applications such as engineering, business activities and industrial designs. The aims of optimization can be very diverse—to minimize the energy consumption and costs, to maximize the profit, outputs, performance and efficiency. Thus, optimization concerns almost every application from engineering design to business planning and from holiday planning to vehicle routing. In reality, because resources, time and money are always limited in real-world applications, the optimal use these valuable resources under various constraints should be sought. As most real-world applications are often highly nonlinear, it requires sophisticated optimization tools so as to find quality and feasible solutions.

Nature-inspired computation is now playing an increasingly important role in many areas, including artificial intelligence, computational intelligence, data mining, machine learning and optimization [1]. Good examples of such algorithms are bat algorithm, cuckoo search, particle swarm optimization, firefly algorithm, bee algorithms and others [1–4]. Obviously, their popularity can be attributed to many factors

X.-S. Yang (✉)
School of Science and Technology, Middlesex University, The Burroughs,
London NW4 4BT, UK
e-mail: x.yang@mdx.ac.uk

X. He
College of Science, Xi'an Polytechnic University, Jinhua South Road,
Xi'an, People's Republic of China

© Springer International Publishing Switzerland 2016
X.-S. Yang (ed.), *Nature-Inspired Computation in Engineering*,
Studies in Computational Intelligence 637, DOI 10.1007/978-3-319-30235-5_1

and one key reason is that these algorithms are simple, flexible, efficient and highly adaptable. In addition, from the implementation point of view, they are very simple to be implemented in any programming language. As a result, these algorithms have been applied in a wide spectrum of problems in real-world applications.

The main objective of this chapter is to provide an overview of the history of the nature-inspired computation and review some of the recent nature-inspired algorithms for optimization. Some applications in engineering will also be highlighted. Therefore, the chapter is organized as follows. Section 2 outlines the basic formulation of optimization problems and the search for optimality, and then Sect. 3 presents the brief history of the recent developments in nature-inspired computation. Section 4 highlights the key features of metaheuristic algorithms, followed by the introduction to some recent algorithms based on swarm intelligence in Sect. 5. Section 6 discusses the choice of algorithms and then Sect. 7 highlights some recent applications of bio-inspired computation in engineering.

2   Optimization and Optimality

Many problems in engineering can be formulated as an optimization problem, though how to formulate and the way of formulations usually depend on the perspective that we are looking at the problem. Though there are many different ways of formulations, the exact formulations may also depend on the subject area and thus such formulations may directly or indirectly affect the choice of the solution techniques. Thus, this may be a subject matter, however, we will focus on the general context of optimization and search for optimality.

2.1   Optimization

From a mathematical point of view, all almost optimization problems can be formulated in the following generic form:

$$\underset{x \in \mathfrak{R}^d}{\text{minimize}} \quad f_i(x), \quad (i = 1, 2, \ldots, M), \tag{1}$$

$$\text{subject to } h_j(x) = 0, \quad (j = 1, 2, \ldots, J), \tag{2}$$

$$g_k(x) \leq 0, \quad (k = 1, 2, \ldots, K), \tag{3}$$

where $f_i(x)$, $h_j(x)$ and $g_k(x)$ are functions of the design vector

$$x = (x_1, x_2, \ldots, x_d)^T. \tag{4}$$

Here the components x_i of x are called design or decision variables, and they can be real continuous, discrete or the mixed (partly continuous and partly discrete). Here, the functions $f_i(x)$ where $i = 1, 2, \ldots, M$ are called the objective functions or simply cost functions, and in the case of $M = 1$, there is only a single objective and the optimization process become a single objective optimization. The space spanned by the decision variables is called the design space or search space $\mathfrak{R}^d$, while the space formed by the objective function values is called the solution space or response space. The equalities for h_j and inequalities for g_k are called constraints.

It is worth pointing out that we can also write the inequalities in the other way ≥ 0, and we can also formulate the objectives as a maximization problem. In a rare but extreme case where there is no objective at all, there are only constraints. Such a problem is called a feasibility problem because any feasible solution is an optimal solution. For some difficult design problems with multiple, complex, potentially conflicting constraints, to find even a feasible solution may be a challenging task.

Obviously, if we try to classify optimization problems according to the number of objectives, then there are two categories: single objective $M = 1$ and multiobjective $M > 1$. Multiobjective optimization is also referred to as multicriteria or even multi-attributes optimization in the literature. In real-world problems, most optimization tasks are multiobjective. Though the algorithms we will discuss in this chapter are equally applicable to multiobjective optimization with some modifications, we will mainly focus on single objective optimization problems.

Similarly, we can also classify optimization in terms of number of constraints $J + K$. If there is no constraint at all (i.e., $J = K = 0$), then it is called an unconstrained optimization problem. If $K = 0$ and $J \geq 1$, it is called an equality-constrained problem, while the case for $J = 0$ and $K \geq 1$ becomes an inequality-constrained problem.

It is worth pointing out that equality constraints have special properties, and require special care. One drawback is that the volume of satisfying an equality is essentially zero in the search space, thus very difficult to get sampling points that satisfy the equality exactly. Some tolerance or allowance is used in practice.

Furthermore, we can also use the actual function forms for classifying optimization problems. The objective functions can be either linear or nonlinear. If the constraints h_j and g_k are all linear, then the problem becomes a linearly constrained optimization problem. If both the constraints and the objective functions are all linear, it becomes a linear programming problem. Here 'programming' has nothing to do with computing programming, it means planning and/or optimization. However, generally speaking, all f_i, h_j and g_k are nonlinear, we have to deal with a nonlinear optimization problem.

2.2 Search for Optimality

Once an optimization problem is formulated correctly, the main task is to find the optimal solutions by a proper solution procedure using the right mathematical techniques. However, there may not be any right solution procedure for some classes of problems, at least, it is not easy to find a right technique to solve a particular type of problem.

In many ways, searching for the optimal solution is very similar to treasure hunting. Imagine a scenario that we are trying to hunt for a hidden treasure in a hilly landscape within a time limit. In one extreme, suppose we are blind-fold without any guidance, the search process is essentially a pure random search, which is usually not efficient. In another extreme, if we are told the treasure is placed at the highest peak of a known region, we will then directly climb up to the steepest cliff and try to reach to the highest peak, and this scenario corresponds to the classical hill-climbing techniques. In most cases, the search is between these extremes. We are not blind-fold, and we do not know where to look for. Obviously, it is physically impossible to search every single square inch of an extremely large hilly region so as to find the treasure.

In reality, the most likely search scenario is that we will do a random walk, while looking for some hints; we look at some place almost randomly, then move to another plausible place, then another and so on. Such a random walk is a main characteristic of modern search algorithms [1]. Obviously, we can either do the treasure-hunting alone, so the whole path is a trajectory-based search, and simulated annealing is such a kind. Alternatively, we can ask a group of people to do the hunting and share the information (and any treasure found), and this scenario uses the so-called swarm intelligence and corresponds to the algorithms such as particle swarm optimization and firefly algorithm, as we will discuss later in detail. If the treasure is really important and if the area is extremely large, the search process will take a very long time. If there is no time limit and if any region is accessible (for example, no islands in a lake), it is theoretically possible to find the ultimate treasure (the global optimal solution).

Going with this analogy even further, we can refine our search strategy a little bit further. Some hunters are better than others. We can only keep the better hunters and recruit new ones, this is something similar to the genetic algorithms or evolutionary algorithms where the search agents are improving. In fact, as we will see in almost all modern metaheuristic algorithms, we try to use the best solutions or agents, and randomize (or replace) the not-so-good ones, while evaluating each individual's competence (fitness) in combination with the system history (use of memory). With such a balance, we intend to design better and efficient optimization algorithms.

3 A Brief History of Nature-Inspired Computation

Humans' approach to problem-solving has always been heuristic or metaheuristic—by trial and error, especially at the early periods of human history. Many important discoveries were done by 'thinking outside the box', and often by accident; that is heuristics. In fact, our daily learning experience during our childhood is dominantly heuristic. Despite its ubiquitous nature, metaheuristics as a scientific method to problem solving is indeed a modern phenomenon, though it is difficult to pinpoint when the metaheuristic method was first used. Alan Turing was probably the first to use heuristic algorithms during the second World War when he was breaking the Enigma ciphers at Bletchley Park. Turing called his search method *heuristic search*, as it could be expected it worked most of time, but there was no guarantee to find the correct solution; however, it was a tremendous success. In his National Physical Laboratory report on *Intelligent machinery* in 1948, Turing outlined his innovative ideas of machine intelligence and learning, neural networks and evolutionary algorithms [7].

The first golden period is the 1960 s and 1970 s for the development of evolutionary algorithms. John Holland and his collaborators at the University of Michigan first developed the genetic algorithms in 1960 s and 1970s. As far back to 1962, Holland studied the adaptive system and was the first to use crossover and recombination manipulations for modeling such system. His seminal book summarizing the development of genetic algorithms was published in 1975 [8]. In the same year, De Jong finished his important dissertation showing the potential and power of genetic algorithms for a wide range of objective functions, either noisy, multimodal or even discontinuous. The essence of a genetic algorithm (GA) is based on the abstraction of Darwinian evolution and natural selection of biological systems and representing them in the mathematical operators: crossover or recombination, mutation, fitness, and selection of the fittest.

During the same decade, Ingo Rechenberg and Hans-Paul Schwefel both then at the Technical University of Berlin developed a search technique for solving optimization problem in aerospace engineering, called evolutionary strategy, in 1963. There was no crossover in this technique, only mutation was used to produce an offspring and an improved solution was kept at each generation. This was essentially a simple trajectory-style hill-climbing algorithm with randomization. On the other hand, as early as 1960, Lawrence J. Fogel intended to use simulated evolution as a learning process as a tool to study artificial intelligence. Then, in 1966, L.J. Fogel, together A.J. Owen and M.J. Walsh, developed the evolutionary programming technique by representing solutions as finite-state machines and randomly mutating one of these machines [9]. The above innovative ideas and methods have evolved into a much wider discipline, called evolutionary algorithms and/or evolutionary computation [10, 11].

It may not be entirely clear how bio-inspired computation is also relevant to artificial neural network, support vector machine, and other many learning techniques in the context of optimization. However, these methods all intend to minimize their learning errors and prediction (capability) errors via iterative trials and errors. For example, artificial neural networks are now routinely used in many applications. In 1943, W. McCulloch and W. Pitts proposed the artificial neurons as simple information processing units. The concept of a neural network was probably first proposed by Alan Turing in his 1948 NPL report concerning 'intelligent machinery' [7, 12]. Significant developments were carried out from the 1940 s and 1950 s to the 1990 s with more than 60 years of history. Another example is the so-called support vector machine as a classification technique that can date back to the earlier work by

V. Vapnik in 1963 on linear classifiers, and the nonlinear classification with kernel techniques were developed by V. Vapnik and his collaborators in the 1990s. A systematical summary in Vapnik's book on the Nature of Statistical Learning Theory was published in 1995 [13].

Another important period for metaheuristic algorithms is the two decades of 1980 s and 1990s. First, the development of simulated annealing (SA) in 1983, an optimization technique, pioneered by S. Kirkpatrick, C.D. Gellat and M.P. Vecchi, inspired by the annealing process of metals. The actual first usage of memory in modern metaheuristics is probably due to Fred Glover's Tabu search in 1986, though his seminal book on Tabu search was published later in 1997 [14]. Marco Dorigo finished his PhD thesis in 1992 on optimization and natural algorithms [2], in which he described his innovative work on ant colony optimization (ACO). This search technique was inspired by the swarm intelligence of social ants using pheromone as a chemical messenger. At the same time in 1992, John R. Koza of Stanford University published a treatise on genetic programming which laid the foundation of a whole new area of machine learning, revolutionizing computer programming [15]. As early as in 1988, Koza applied his first patent on genetic programming. The basic idea is to use the genetic principle to breed computer programs so as to gradually produce the best programs for a given type of problem.

In 1995, the particle swarm optimization (PSO) was developed by American social psychologist James Kennedy, and engineer Russell C. Eberhart [3], based on the swarming behaviour of fish and birds. The multiple agents, called particles, swarm around the search space starting from some initial random guess. The swarm communicates the current best and shares the global best so as to focus on the quality solutions. Since its development, there have been about 20 different variants of particle swarm optimization techniques, and have been applied to almost all areas of tough optimization problems.

Then, slightly later in 1996 and 1997, R. Storn and K. Price developed their vector-based evolutionary algorithm, called differential evolution (DE) [16], and this algorithm proves more efficient than genetic algorithms in many applications. In 1997, the 'no free lunch theorems for optimization' were proved by D.H. Wolpert and W.G. Macready [17, 18]. Researchers have been always trying to find better algorithms, or even universally robust algorithms, for optimization, especially for tough NP-hard optimization problems. However, these theorems state that if algorithm A performs better than algorithm B for some optimization functions, then B will outperform A for other functions. That is to say, if averaged over all possible function space, both algorithms A and B will perform on average equally well. Alternatively, there is no universally better algorithms exist. However, researchers realized that we do not need the average over all possible functions for a given optimization problem. What we want is to find the best solutions, which has nothing to do with average over all possible function space. In addition, we can accept the fact that there is no universal or magical tool, but we do know from our experience that some algorithms indeed outperform others for given types of optimization problems. So the research may now focus on finding the best and most efficient algorithm(s) for a given set of problems. The objective is to design better algorithms for most types of problems, not for all the problems. Therefore, the search is still on.

Another very exciting period for developing metaheuristic algorithms is the first decade of the 21st century. In 2001, Zong Woo Geem et al. developed the harmony search (HS) algorithm [19], which has been widely applied in solving various optimization problems such as water distribution, transport modelling and scheduling. In 2004, S. Nakrani and C. Tovey proposed the honey bee algorithm and its application for optimizing Internet hosting centers [20], which followed by the development of virtual bee algorithm by Xin-She Yang in 2005. At the same time, the bees algorithm was developed by D.T. Pham et al. in 2005 and the artificial bee colony (ABC) was developed by D. Karaboga in 2005. Then, in late 2007 and early 2008, the firefly algorithm (FA) was developed by Xin-She Yang [1, 4], which has generated a wide range of interests. In 2009, Xin-She Yang at Cambridge University, UK, and Suash Deb at Raman College of Engineering, India, proposed an efficient cuckoo search (CS) algorithm [21, 22], and it has been demonstrated that CS can be far more effective than some existing metaheuristic algorithms. In 2010, the bat algorithm was developed by Xin-She Yang for continuous optimization, based on the echolocation behaviour of microbats [23]. In 2012, the flower pollination algorithm was developed by Xin-She Yang, and its efficiency is very promising.

As the literature is expanding rapidly, the number of nature-inspired algorithms has increased dramatically [1, 24] As we can see, more and more metaheuristic algorithms are being developed. Such a diverse range of algorithms necessitates a systematic summary of various metaheuristic algorithms, however, we only briefly outline a few of these recent algorithms in the rest of the chapter. But before we proceed, let us pause and highlight the key characteristics of nature-inspired algorithms.

4 Main Characteristics of Nature-Inspired Algorithms

The key characteristics of nature-inspired optimization algorithms can be analyzed from many different perspectives. From the algorithmic development point of view, we can analyze algorithms in terms of algorithm search behaviour, adaptation and diverse as well as algorithmic components in terms of mathematical equations for iterations.

4.1 Exploration and Exploitation

The key aim of an algorithm is to generate new solutions that should be better than previous solutions. For an ideal algorithm, new solutions should be always better than existing solutions and it can be expected that the most efficient algorithms are to find the best solutions with the least minimum efforts (ideally in one move). However, such ideal algorithms may not exist at all.

For stochastic algorithms, solutions do not always get better. In fact, it can be advantageous to select not-so-good solutions, which can help the search process escape from being trapped at any local optima. Though this may be counter-intuitive, such stochastic nature now forms the essential component of modern metaheuristic algorithms [4]. Exploitation typically uses any information obtained from the problem of interest so as to help to generate new solutions that are better than existing solutions. However, this process is typically local, and information (such as gradients) is also local. Therefore, it is mainly for local search. For example, hill-climbing is a method that uses derivative information to guide the search procedure. In fact, new steps always try to climb up the local gradient. The advantage of exploitation is that it usually leads to very high convergence rates, but its disadvantage is that it can get stuck in a local optimum because the final solution point largely depends on the starting point.

On the other hand, exploration makes it possible to explore the search space in far away regions more efficiently [25], and it can generate solutions with enough diversity and far from the current solutions. Therefore, the search is typically on a global scale. The advantage of exploration is that it is less likely to get stuck in a local mode, and the global optimality can be more accessible. However, its disadvantages are slow convergence and waste of lot computational efforts because many new solutions can be far from global optimality.

However, whether local search or global search, all depends on the actual search mechanisms within an algorithm. Sometimes, there is no clear cut between local or global. In addition, a fine balance may be required so that an algorithm can achieve the good performance. Too much exploitation and too little exploration means the system may converge more quickly, but the probability of finding the true global optimality may be low. Conversely, too little exploitation and too much exploration can cause the search path meander with very slow convergence. The optimal balance should mean the right amount of exploration and exploitation, which may lead to the optimal performance of an algorithm. Therefore, a proper balance is crucially important to ensure the good performance of an algorithm.

4.2   Diversity and Adaptation

Looking from a different perspective, nature-inspired algorithms can have both diversity and adaptation. Adaptation in nature-inspired algorithms can take many forms. For example, the ways to balance exploration and exploitation are the key form of adaptation [26]. As diversity can be intrinsically linked with adaptation, it is better not to discuss these two features separately. For example, in genetic algorithms, representations of solutions are usually in binary or real-valued strings [8, 26], while in swarm-intelligence-based algorithms, representations mostly use real number solution vectors. As another example, the population size used in an algorithm can be fixed or varying. Adaptation in this case may mean to vary the population size so as to maximize the overall performance.

For a given algorithm, adaptation can also occur to adjust its algorithm-dependent parameters. As the performance of an algorithm can largely depend on its parameters, the choice of these parameter values can be very important. Similarly, diversity in metaheuristic algorithms can also take many forms. The simplest diversity is to allow the variations of solutions in the population by randomization. For example, solution diversity in genetic algorithms is mainly controlled by the mutation rate and crossover mechanisms, while in simulated annealing, diversity is achieved by random walks. In addition, adaptation can also be in the form of self-tuning in terms of parameters so as to achieve better performance automatically [27].

In most swarm-intelligence-based algorithms, new solutions are generated according to a set of deterministic equations, which also include some random variables. Diversity is represented by the variations, often in terms of the population variance. Once the population variance is getting smaller (approaching zero), diversity also decreases, leading to converged solution sets. However, if diversity is reduced too quickly, premature convergence may occur. Therefore, a right amount of randomness and the right form of randomization can be crucial.

Both diversity and adaptation are important to ensure the effectiveness of an algorithm. A good diversity will ensure the population can explore different regions and can thus maintain a non-zero probability of finding the global optimality of the problem. In addition, good adaption will enable the algorithm to adapt to suit the problem and landscape under consideration, and thus may ensure a potentially better convergence than non-adaptive approaches. However, it is not yet clear what a good degree of diversity should be and what kind of adaptation mechanisms can be used.

4.3 Key Operators in Algorithms

Algorithms can also be analyzed by studying in detail the key algorithmic operators used in the construction of the algorithms. For example, in the well-established class of genetic algorithms [8], genetic operators such as crossover (or recombination), mutation and selection are used [26].

In genetic algorithms, crossover is the operation of generating two new solutions (offsprings) from two existing solutions (parents) by swapping relevant/corresponding parts of their solutions. This is similar to the main crossover feature in the biological systems. Crossover usually provides good mixing of solution characteristics and can usually generate completely new solutions if the two parents are different. Obviously, when the two parents are identical, offspring solutions will also be identical, and thus provides a good mechanism to maintain good convergence. It is worth pointing out that crossover in contemporary algorithms may take different forms, though its essence remains the same.

Mutation is a mechanism to generate a new solution from a single solution by changing a single site or multiple sites. As in the evolution in nature, mutation often generates new characteristics that can adapt to new environments, and thus new solutions during the search process can also be generated in this way. However, in many nature-inspired algorithms, mutation typically becomes a vector and thus has taken very different forms from the original binary mutation in the original genetic algorithms.

Though both crossover and mutation are ways of generating new solutions, selection provides a pressure for evolution. In other words, selection provides a measure or mechanism to determine what is better and selects the fittest. This mimics the key feature of the Darwinian evolution in terms of the survival of the fittest. Without selection, new solutions and new characteristics will not be selected properly, which may lead to a diverse and less convergent system. With too much selection pressure, many new characteristics will die away quickly. If the environment also changes and if some solutions/characteristics dominate the population, the system may lead to premature convergence. Thus, a proper selection pressure is also important to ensure that only good characteristics or solutions that happen to fit into the new environment can survive.

If we try to link crossover and mutation with exploration and exploitation, we have to look at them in greater detail because such links can be very subtle. For example, crossover can provide both exploration and exploitation capabilities, while mutation mainly provides exploration. On the other hand, selection provides a good way of exploitation by selecting good solutions. Thus, in order to provide a good exploration ability to an algorithm, a higher crossover rate is needed, and that is why the crossover probability is typically over 0.9 in genetic algorithms. However, new solutions should not be too far from existing solutions in many cases, and to avoid too much exploration, the mutation rate should be usually low. For example, in genetic algorithms, the mutation rate is typically under 0.05 [4]. It is worth pointing out that these values are based on empirical observations (both numerically and biologically). In different metaheuristic algorithms, it is usually quite a challenging task to decide what values are most appropriate, and such choices may need detailed parametric studies and numerical experiments.

5   Some Swarm-Based Algorithms

As nature-inspired algorithms have become more popular, the literature has expanded significantly. In fact, the number of nature-inspired algorithms has increased dramatically in recent years, thus it is not possible to review a good subset of all these algorithms here. To demonstrate the main points, we will only briefly introduce a few algorithms that are among most recent and most popular algorithms.

All almost algorithms use some updating rules, derived from some characteristics in nature, though a vast majority draw inspiration from swarm-based characteristics, and thus forming a class of swarm intelligence. It is worth pointing out that the updating or iterative equations can be either linear or nonlinear, though most algorithms have linear updating equations. Linear systems may be easier to implement, but the diversity and richness of the system behaviour may be limited.

On the other hand, for the algorithms with nonlinear updating systems, the characteristics of the algorithm can be richer, which may lead to some advantages over algorithms with linear updating equations. For example, studies show that firefly algorithm can automatically subdivide the whole population into multiple sub-swarms due to its nonlinear distance-dependent attraction mechanism. At the moment, it is still not clear if nonlinear systems are always potentially better. In the rest of this section, when we describe the key formulations of each algorithm, we will mainly focus on the key characteristics of its updating equations as an iterative system.

5.1 Particle Swarm Optimization

Particle swarm optimization (PSO), developed by Kennedy and Eberhart in 1995 [3] has two updating/iterative equations for position x_i and velocity v_i of a particle system:

$$v_i^{t+1} = v_i^t + \alpha \epsilon_1 [g^* - x_i^t] + \beta \epsilon_2 [x_i^* - x_i^t], \tag{5}$$

$$x_i^{t+1} = x_i^t + v_i^{t+1}. \tag{6}$$

There are two parameters α and β in addition to the population size n. Here, two random vectors ϵ_1 and ϵ_2 are uniformly distributed in [0, 1].

As this system is linear, it is quite straightforward to carry out stability analysis [28]. Though the PSO system usually converges very quickly, it can have premature convergence. To remedy this, over 20 different variants have been proposed with moderate success. Among many improvements, the use of an inertia function by Shi and Eberhart [29] seems to stabilize the system well. Other developments include the accelerated PSO by Yang et al. [30] and some reasoning techniques with PSO by Fister Jr. et al. [31]. Hybridization with other algorithms also proves useful.

5.2 Firefly Algorithm

The firefly algorithm (FA) developed by Xin-She Yang in 2008 [4] has a single nonlinear equation for updating the locations (or solutions) of fireflies:

$$x_i^{t+1} = x_i^t + \beta_0 e^{-\gamma r_{ij}^2} (x_j^t - x_i^t) + \alpha \, \epsilon_i^t, \tag{7}$$

where the second term between any two fireflies (i and j) due to the attraction is highly nonlinear because the attraction is distance-dependent. Here, β_0 is the attractiveness at $r = 0$, while α is the randomization parameter. In addition, ϵ_i^t is a vector of random numbers drawn from a Gaussian distribution at time t. The nonlinear nature of the updating equation means that the local short-distance attraction is much stronger than the long-range attraction, and consequently, the whole population can automatically subdivide into multiple subgroups (or multi-swarms). Under the right conditions, each subgroup can swarm around a local mode, thus, FA can naturally deal with multimodal problems effectively.

Studies show that firefly algorithm can be very efficient in solving many different problems such as classifications and clustering problems [32] as well as scheduling problems [33]. The extensive studies of the firefly algorithm and its variants were reviewed by Fister et al. [24, 34, 35].

5.3   Cuckoo Search

Cuckoo search (CS), developed in 2009 by Xin-She Yang and Suash Deb [21], is another nonlinear system [22, 36], enhanced by Lévy flights [37]. This makes CS potentially more efficient than PSO and genetic algorithms.

The nonlinear equations are also controlled by a switching parameter or probability p_a. The nonlinear system can be written as

$$x_i^{t+1} = x_i^t + \alpha s \otimes H(p_a - \epsilon) \otimes (x_j^t - x_k^t), \tag{8}$$

$$x_i^{t+1} = x_i^t + \alpha L(s, \lambda), \tag{9}$$

where two different solutions x_j^t and x_k^t are randomly selected. In addition, $H(u)$ is a Heaviside step function, ϵ is a random number drawn from a uniform distribution, and s is the step size. The so-called Lévy flights are realized by drawing random step sizes $L(s, \lambda)$ from a Lévy distribution, which can be written as

$$L(s, \lambda) \sim \frac{\lambda \Gamma(\lambda) \sin(\pi \lambda / 2)}{\pi} \frac{1}{s^{1+\lambda}}, \quad (s > 0), \tag{10}$$

where $\sim$ denotes the drawing of samples from a probability distribution. In addition, $\alpha > 0$ is the step size scaling factor, which should be related to the scales of the problem of interest.

Some recent reviews have been carried out by Yang and Deb [38] and Fister Jr. et al. [39]. Mathematical analysis also suggested that CS can have global convergence [40]. Recent studies also suggests that CS can have autozooming capabilities so that the search process can automatically focus on the promising areas due to the combination of self-similar, fractal-like and multiscale search capabilities [1]. By analyzing the algorithmic system, it can be seen that simulated annealing, differential evolution and APSO are special cases of CS, and that is one of the reasons why CS is so efficient.

5.4 Bat Algorithm

The bat algorithm, developed by Xin-She Yang in 2010 [23] based on the echolocation behavior of microbats, is a linear system with updating equations for velocity v_i^t and location x_i^t of a bat. At iteration t in a d-dimensional search or solution space, the linear system can be written as

$$f_i = f_{\min} + (f_{\max} - f_{\min})\beta, \tag{11}$$

$$v_i^t = v_i^{t-1} + (x_i^{t-1} - x_*)f_i, \tag{12}$$

$$x_i^t = x_i^{t-1} + v_i^t, \tag{13}$$

where x_* is the current best solution and $\beta \in [0, 1]$ is a random vector drawn from a uniform distribution.

Though variations of the loudness and pulse emission rates are regulated

$$A_i^{t+1} = \alpha A_i^t, \quad r_i^{t+1} = r_i^0[1 - \exp(-\gamma t)], \tag{14}$$

they seems to control the exploration and exploitation components in the algorithm. Here, $0 < \alpha < 1$ and $\gamma > 0$ are constants. BA has attracted a lot of interest and thus the literature is expanding. For example, Yang extended it to multiobjective optimization [41, 42] and Fister et al. formulated a hybrid bat algorithm [43, 44].

5.5 Differential Evolution

Differential evolution (DE), developed by R. Storn and K. Price in 1996 and 1997 [16, 45], uses a linear mutation equation

$$u_i^t = x_r^t + F(x_p^t - x_q^t), \tag{15}$$

where F is the differential weight in the range of $[0, 2]$. Here, r, p, q, i are four different integers generated by random permutation.

The crossover operator in DE is controlled by a crossover probability $C_r \in [0, 1]$ and the actual crossover can be carried out in two ways: binomial and exponential.

Crossover can typically be carried out along each dimension $j = 1, 1, \ldots, d$ where d is the number of dimensions. Thus, we have

$$
x_{j,i}^{t+1} =
\begin{cases}
u_{j,i}^t & \text{if rand} < C_r \text{ or } j = I, \\
\\
x_{j,i}^t & \text{if rand} > C_r \text{ and } j \neq I,
\end{cases}
\tag{16}
$$

where I is a random integer from 1 to d, so that $u_i^{t+1} \neq x_i^t$. It may not easy to figure out whether such crossover is linear or not, a detailed analysis indicates that crossover can be treated as a linear operator and thus DE is still a linear system in terms of updating equations. There are many different variants of DE [46] and many hybridized version in the literature.

5.6   Flower Pollination Algorithm

Another linear system is the flower pollination algorithm (FPA), developed by Xin-She Yang in 2012 [47], inspired by the flower pollination process of flowering plants. The main updating equation for a solution/position x_i at any iteration t is

$$
x_i^{t+1} = x_i^t + \gamma L(\lambda)(g_* - x_i^t),
\tag{17}
$$

where g_* is the current best solution found among all solutions and γ is a scaling factor to control the step size. The step sizes as Lévy flights are drawn from a Levy distribution, that is

$$
L(\lambda) \sim \frac{\lambda \Gamma(\lambda) \sin(\pi \lambda / 2)}{\pi} \frac{1}{s^{1+\lambda}}, \quad (s \gg s_0 > 0).
\tag{18}
$$

Here $\Gamma(\lambda)$ is the standard gamma function, and this distribution is valid for large steps $s > 0$.

Another equation is also linear

$$
x_i^{t+1} = x_i^t + \epsilon(x_j^t - x_k^t),
\tag{19}
$$

which is mainly for local search, though whether it is local or global in this equation will depend on the solutions chosen from the population. If two solutions are very different or far away, the search can be global, while two very similar solutions will typically lead to local search. Here, x_j^t and x_k^t are pollen from different flowers of the same plant species. Here, ϵ from a uniform distribution in $[0, 1]$. In essence, this is mostly a local mutation and mixing step, which can help to converge in a subspace. However, it is worth pointing out that the generation of new moves using Lévy flights can be tricky. There are efficient ways to draw random steps correctly from a Lévy distribution [1, 37]. The switch between two search branches are controlled by a probability p_s. Recent studies suggested that flower pollination algorithm is very efficient for multiobjective optimization [48].

6 Theoretical Analysis and Choice of Algorithms

Though different algorithms can have very different iterative equations, it is still possible to find some common features among different algorithms. However, to systematically analyze many different algorithms can be a very challenging task.

On the other hand, there is no free lunch in theory when the performance metrics averaged over all possible problems [17], but ranking of algorithms may still be possible for a given set of particular problems because the performance in this case is no longer concerned with all problems and averaging. Thus, for a given set or type of problems, one of the main tasks is to find some effective algorithms to use among many different algorithms and thus the choice of algorithms may not be an easy task.

6.1 Theoretical Aspects of Algorithms

The ways of generating new solutions via updating equations depend on the structure and topology of the search neighbourhood. New solution can be in either a local neighborhood such as local random walks, the global domain such as those moves by Lévy or uniform initialization, or cross-scale (both local and global, e.g., Lévy flights, exponential, power-law, heavy-tailed). For simplicity, let us focus on the most fundamental operators such as crossover, mutation and selection. From the mathematical point of view, selection is easy to understand, we will now focus only on crossover and mutation.

The crossover operator can be written mathematically as

$$\begin{pmatrix} x_i^{t+1} \\ x_j^{t+1} \end{pmatrix} = C(x_i^t, x_j^t, p_c) = C(p_c) \begin{pmatrix} x_i^t \\ x_j^t \end{pmatrix}, \tag{20}$$

where p_c is the crossover probability, though the exact form of $C(\ )$ depends on the actual crossover manipulations such as at one site or at multiple sites simultaneously. The selection of i and j can be by random permutation. Furthermore, the choice of parents can often be fitness-dependent, based on the relative fitness of the parents in the population. In this particular case, the functional form for the crossover function can be even more complex. For example, $C(\ )$ can depend on all the individuals in the population, which may lead to $C(x_1^t, x_2^t, \ldots, x_n^t, p_c)$ where n is the population size.

On the other hand, mutation is the variation or modification of an existing solution and can be written schematically as

$$x_i^{t+1} = M(x_i^t, p_m), \tag{21}$$

where p_m is the mutation rate. However, the form $M(\ )$ depends on the coding and the number of mutation sites. This can be written in most cases as a random walk

$$x_i^{t+1} \text{(new solution)} = x_i^t \text{(old solution)} + \alpha \text{(randomization)}, \tag{22}$$

where α is a scaling factor controlling how far the random walks can go [4, 23]. However, mutation can also be carried out over a subset of the population or the mutation operator can also be affected by more than one solution. For example, the mutation operator in differential evolution takes the form $x_r^t + F(x_p^t - x_q^t)$, which involves three different solutions.

In general, the solutions can be generated in parallel by random permutation, and thus we may have a more generic form for modifications

$$\begin{pmatrix} x_1^{t+1} \\ \vdots \\ x_n^{t+1} \end{pmatrix} = G(x_1^t, x_2^t, \ldots, x_n^t, \varepsilon, \beta), \tag{23}$$

where $G(\ )$ can be in a very complex form, involving vectors, matrices, random variable ε and parameter β. However, it still lacks a unified mathematical framework for algorithm analysis.

6.2   Choice of Algorithms

A very important but also very practical question is how to choose an algorithm for a given problem. This may be implicitly linked to another question: what type of problems can an algorithm solve most effectively. In many applications, the problem under consideration seem to be fixed, we have to use the right tool or methods to solve in the most effective way. Therefore, there are two types of choices and thus two relevant questions:

- For a given type of problems, what is the best algorithm to use?
- For a given algorithm, what kinds of problems can it solve?

The first question is harder than the second question, though the latter is not easy to answer either. For a given type of problems, there may be a set of efficient algorithms to solve such problems. However, in many cases, we may not know how efficient an algorithm can be before we actually try it. In some cases, such algorithms may still need to be developed. Even for existing algorithms, the choice largely depends on the expertise of the decision-maker, the resources and the type of problems. Ideally, the best available algorithms and tools should be used to solve a given problem; however, the proper use of these tools may still depend on the experience of the user.

In addition, the resources such as computational costs and software availability and time allowed to produce the solution will also be important factors in deciding what algorithms and methods to use.

On the other hand, for a given algorithm, the type of problems it can solve can be explored by using it to solve various kinds of problems and then compare and rank so as to find out how efficient it may be. In this way, the advantages and disadvantages can be identified, and such knowledge can be used to guide the choice of algorithm(s) and the type of problems to tackle. The good thing is that the majority of the literature, including hundreds of books, have placed tremendous emphasis in answering this question. Therefore, for traditional algorithms such as gradient-based algorithms and simplex methods, we know what types of problems they usually can solve. However, for new algorithms, as in the cases of most nature-inspired algorithms, we have to carry out extensive studies to validate and test their performance. Obviously, any specific knowledge about a particular problem is always helpful for the appropriate choice of the best and most efficient methods for the optimization From the algorithm development point of view, how to best incorporate problem-specific knowledge is still an ongoing challenging question.

7 Applications in Engineering

The applications of nature-inspired algorithms in engineering is very diverse and to cover a good fraction of these applications may take a big volume. In the present book, some of the latest developments will be reviewed and introduced.

Therefore, in the rest of this section, we will briefly highlight a few applications to show the diversity and effectiveness of nature-inspired optimization algorithms.

- *Structural optimization*: Design optimization in civil engineering is essentially structural optimization because the main tasks to design structures so as to maximize the performance index and to minimize the costs, subject to complex stringent design codes. Typical examples are pressure vessel design, speed reducer design, dome and tower designs [22, 41, 42], as well as combination of optimization with finite element simulation and other designs [42, 49].
- *Scheduling and Routing*: Scheduling problems such as airline scheduling and vehicle routing can have important real-world applications. The well-known example is the so-called travelling salesman problem that requires to visit each city exactly once so that the overall travelled distance must be minimized. Nature-inspired algorithms have been used to solve such tough problems with promising results [50–52].
- *Software testing*: In software engineering, a substantial amount of the costs for software developments can be related to software testing to make sure the

software can execute smoothly and to meet the design requirements. This means that multiple independent test paths should be generated to test the behaviour of the software, though the generation of truly independent paths can be a difficult task. Nature-inspired optimization algorithms have also been used in this area with good results [53].

- *Image processing*: Image processing is a big area with a huge literature. Image segmentation and feature selection can often be formulated as an optimization problem, and thus can be tackled by optimization techniques in combination with traditional image processing techniques [54]. Image automatic registration and clustering can be solved using nature-inspired algorithms with good performance [32, 55].
- *Data mining*: Data mining is an active research area with diverse applications. Classification and clustering can be closely related to optimization and many hybrid techniques have recently been developed by combing traditional data mining methods such as k-mean clustering with nature-inspired optimization algorithms such as the firefly algorithm, cuckoo search and bat algorithm. Recent studies by Fong et al. showed that such hybrid methods can obtain very good results [56]. For example, Senthilnath et al. compared over a dozen different clustering algorithms, they concluded that the approach based the firefly algorithm can obtain the best results with the least amount of computational efforts [32].

Obviously, there are many other applications and case studies, interested readers can refer to more specialized literature [1, 5, 6, 57] and the later chapters of this book

References

1. Yang, X.S.: Nature-Inspired Optimization Algorithms. Elsevier, London (2014)
2. Dorigo, M., Di Caro, G., Gambardella, L.M.: Ant algorithms for discrite optimization. Artif. Life **5**(2), 137–172 (1999)
3. Kennedy, J., Eberhart, R.C.: Particle swarm optimization. In: Proceedings of IEEE International Conference on Neural Networks, pp. 1942–1948. Piscataway, NJ (1995)
4. Yang, X.S.: Nat.-Inspir. Metaheuristic Algorithms. Luniver Press, Bristol (2008)
5. Eiben, A.E., Smit, S.K.: Parameter tuning for configuring and analyzing evolutionary algorithms. Swarm Evol. Comput. **1**(1), 19–31 (2011)
6. Passino, K.M.: Bactorial foraging optimization. Int. J. Swarm Intell. Res. **1**(1), 1–16 (2010)
7. Copeland, B.J.: The Essential Turing. Oxford University Press, Oxford (2004)
8. Holland, J.: Adaptation in Natural and Artificial Systems. University of Michigan Press, Ann Anbor (1975)
9. Fogel, L.J., Owens, A.J., Walsh, M.J.: Artificial Intelligence Through Simulated Evolution. Wiley, New York (1966)
10. Judea, P.: Heuristics. Addison-Wesley, New York (1984)
11. Schrijver, A.: On the history of combinatorial optimization (till 1960). In: Aardal, K., Nemhauser, G.L., Weismantel, R. (eds.) Handbook of Discrete Optimization, pp. 1–68. Elsevier, Amsterdam (2005)
12. Turing, A.M.: Intelligent Machinery. National Physical Laboratory, Technical report (1948)
13. Vapnik, V.: Nat. Stat. Learn. Theory. Springer, New York (1995)

14. Glover, F., Laguna, M.: Tabu Search. Kluwer Academic Publishers, Boston (1997)
15. Koza, J.R.: Genetic Programming: one the Programming of Computers by Means of Natural Selection. MIT Press, MA (1992)
16. Storn, R., Price, K.: Differential evolution—a simple and efficient heuristic for global optimization over continuous spaces. J. Glob. Optim. **11**(4), 341–359 (1997)
17. Wolpert, D.H., Macready, W.G.: No free lunch theorems for optimization. IEEE Trans. Evol. Comput. **1**(1), 67–82 (1997)
18. Wolpert, D.H., Macready, W.G.: Coevolutonary free lunches. IEEE Trans. Evol. Comput. **9**(6), 721–735 (2005)
19. Geem, Z.W., Kim, J.H., Loganathan, G.V.: A new heuristic optimization: harmony search. Simulation **76**(2), 60–68 (2001)
20. Nakrani, S., Tovey, C.: On honey bees and dynamic server allocation in Internet hostubg centers. Adapt. Behav. **12**(3), 223–240 (2004)
21. Yang, X.S., Deb, S.: Cuckoo search via Lévy flights. In: Proceedings of World Congress on Nature and Biologically Inspired Computing (NaBIC 2009), pp. 210–214. IEEE Publications, USA (2009)
22. Yang, X.S., Deb, S.: Engineering optimization by cuckoo search. Int. J. Math. Model. Numer. Optisation **1**(4), 330–343 (2010)
23. Yang, X.S.: A new metaheuristic bat-inspired algorithm. In: Nature Inspired Cooperative Strategies for Optimisation (NICSO 2010). Springer, Studies in Computational Intelligence, vol. 284, pp. 65–74 (2010)
24. Fister, I., Fister Jr., I., Yang, X.S., Brest, J.: A comprehensive review of firefly algorithms. Swarm Evol. Comput. **13**(1), 34–46 (2013)
25. Blum, C., Roli, A.: Metaheuristics in combinatorial optimisation: overview and conceptual comparision. ACM Comput. Surv. **35**, 268–308 (2003)
26. Booker, L., Forrest, S., Mitchell, M., Riolo, R.: Perspectives on Adaptation in Natural and Artificial Systems. Oxford University Press, Oxford (2005)
27. Yang, X.S., Deb, S., Loomes, M., Karamanoglu, M.: A framework for self-tuning optimization algorithm. Neural Comput. Appl. **23**(7–8), 2051–2057 (2013)
28. Clerc, M., Kennedy, J.: The particle swarm—explosion, stability, and convergence in a multi-dimensional complex space. IEEE Trans. Evol. Comput. **6**(1), 58–73 (2002)
29. Shi, Y.H., Eberhart, R.: A modified particle swarm optimizer. In: Proceedings of the 1998 IEEE World Congress on Computational Intelligence, 4–9 May 1998. IEEE Press, Anchorage, pp. 69–73 (1998)
30. Yang, X.S., Deb, S., Fong, S.: Accelerated particle swarm optimization and support vector machine for business optimization and applications. In: Networked Digital Technologies 2011, Communications in Computer and Information Science, vol. 136, pp. 53–66 (2011)
31. Fister Jr., I., Yang, X.S., Ljubič, K., Fister, D., Brest, J., Fister, I.: Towards the novel reasoning among particles in PSO by the use of RDF and SPARQL. Sci. World J. **2014**, article ID. 121782, (2014). http://dx.doi.org/10.1155/2014/121782
32. Senthilnath, J., Omkar, S.N., Mani, V.: Clustering using firely algorithm: performance study. Swarm Evol. Comput. **1**(3), 164–171 (2011)
33. Yousif, A., Abdullah, A.H., Nor, S.M., Abdelaziz, A.A.: Scheduling jobs on grid computing using firefly algorithm. J. Theor. Appl. Inform. Technol. **33**(2), 155–164 (2011)
34. Fister, I., Yang, X.S., Fister, D., Fister Jr., I.: Firefly algorithm: a brief review of the expanding literature. In: Cuckoo Search and Firefly Algorithm: Theory and Applications, Studies in Computational Intelligence, vol. 516, pp. 347–360. Springer, Heidelberg (2014)
35. Fister, I., Yang, X.-S., Brest, J., Fister Jr., I.: Modified firefly algorithm using quaternion representation. Expert Syst. Appl. **40**(18), 7220–7230 (2013)
36. Yang, X.S., Deb, S.: Multiobjective cuckoo search for design optimization. Compute. Oper. Res. **40**(6), 1616–1624 (2013)
37. Pavlyukevich, I.: Lévy flights, non-local search and simulated annealing. J. Comput. Phys. **226**(12), 1830–1844 (2007)

38. Yang, X.S., Deb, S.: Cuckoo search: recent advances and applications. Neural Comput. Appl. **24**(1), 169–174 (2014)
39. Fister Jr., I., Yang, X.S., Fister, D., Fister, I.: Cuckoo search: a brief literature review. In: Cuckoo Search and Firefly Algorithm: Theory and Applications, Studies in Computational Intelligence, vol. 516, pp. 49–62. Springer, Heidelberg (2014)
40. Wang, F., He, X.S., Wang, Y., Yang, S.M.: Markov model and convergence analysis based on cuckoo search algorithm. Comput. Eng. **38**(11), 180–185 (2012) (in Chinese)
41. Yang, X.S.: Bat algorithm for multi-objective optimisation. Int. J. Bio-Inspir. Comput. **3**(5), 267–274 (2011)
42. Yang, X.S., Gandomi, A.H.: Bat algorithm: a novel approach for global engineering optimization. Eng. Comput. **29**(5), 1–18 (2012)
43. Fister Jr. I., Fong, S., Brest, J., Fister, I.: A novel hybrid self-adaptive bat algorithm. Sci. World J. **2014**, article ID 709738 (2014). http://dx.doi.org/10.1155/2014/709738
44. Fister Jr., I., Fister, D., Yang, X.S.: A hybrid bat algorithm. Elektrotehniski Vestn. **80**(1–2), 1–7 (2013)
45. Storn, R.: On the usage of differential evolution for function optimization. Biennial Conference of the North American Fuzzy Information Processing Society (NAFIPS). Berkeley, CA **1996**, 519–523 (1996)
46. Price, K., Storn, R., Lampinen, J.: Differential Evolution: a Practical Approach to Global Optimization. Springer, Berlin (2005)
47. Yang, X.S.: Flower pollination algorithm for global optimization. In: Unconventional Computation and Natural Computation, pp. 240–249. Springer (2012)
48. Yang, X.S., Karamanoglu, M., He, X.S.: Flower pollination algorithm: a novel approach for multiobjective optimization. Eng. Optim. **46**(9), 1222–1237 (2014)
49. Bekdas, G., Nigdeli, S.M., Yang, X.S.: Sizing optimization of truss structures using flower pollination algorithm. Appl. Soft Comput. **37**(1), 322–331 (2015)
50. Marichelvam, M.K., Prahaharan, T., Yang, X.S.: Improved cuckoo search algorithm for hybrid flow shop scheduling problems to minimize makespan. Appl. Soft Comput. **19**(1), 93–101 (2014)
51. Ouaarab, A., Ahiod, B., Yang, X.S.: Discrete cuckoo search algorithm for the travelling salesman problem. Neural Comput. Appl. **24**(7–8), 1659–1669 (2014)
52. Ouaarab, A., Ahiod, B., Yang, X.S.: Random-key cuckoo search for the travelling salesman problem. Soft. Comput. **19**(4), 1099–1106 (2015)
53. Srivastava, P.R., Millikarjun, B., Yang, X.S.: Optimal test sequence generation using firefly algorithm. Swarm Evol. Comput. **8**(1), 44–53 (2013)
54. Nandy, S., Yang, X.S., Sarkar, P.P., Das, A.: Color image segmentation by cuckoo search. Intell. Autom. Soft Comput. **21**(4), 673–685 (2015)
55. Senthilnath, J., Yang, X.S., Benediktsson, J.A.: Automatic registration of multi-temporal remote sensing images based on nature-inspired techniques. Int. J. Image Data Fusion **5**(4), 263–284 (2014)
56. Fong, S., Deb, S., Yang, X.S., Li, J.Y.: Metaheuristic swarm search for feature selection in life science classificaiton. IEEE IT Prof. **16**(4), 24–29 (2014)
57. Yang, X.S.: Recent advances in swarm intelligence and evolutionary computation. In: Studies in Computational Intelligence, vol. 585. Springer (2015)

An Evolutionary Discrete Firefly Algorithm with Novel Operators for Solving the Vehicle Routing Problem with Time Windows

Eneko Osaba, Roberto Carballedo, Xin-She Yang and Fernando Diaz

Abstract An evolutionary discrete version of the Firefly Algorithm (EDFA) is presented in this chapter for solving the well-known Vehicle Routing Problem with Time Windows (VRPTW). The contribution of this work is not only the adaptation of the EDFA to the VRPTW, but also with some novel route optimization operators. These operators incorporate the process of minimizing the number of routes for a solution in the search process where node selective extractions and subsequent reinsertion are performed. The new operators analyze all routes of the current solution and thus increase the diversification capacity of the search process (in contrast with the traditional node and arc exchange based operators). With the aim of proving that the proposed EDFA and operators are effective, some different versions of the EDFA are compared. The present work includes the experimentation with all the 56 instances of the well-known VRPTW set. In order to obtain rigorous and fair conclusions, two different statistical tests have been conducted.

Keywords Firefly Algorithm · Discrete Firefly Algorithm · Vehicle Routing Problem with Time Windows · Traveling Salesman Problem · Combinatorial optimization

E. Osaba (✉) · R. Carballedo · F. Diaz
Deusto Institute of Technology (DeustoTech), University of Deusto,
Av. Universidades 24, 48007 Bilbao, Spain
e-mail: e.osaba@deusto.es

R. Carballedo
e-mail: roberto.carballedo@deusto.es

F. Diaz
e-mail: fernando.diaz@deusto.es

X.-S. Yang
School of Science and Technology, Middlesex University,
Hendon Campus, London NW4 4BT, UK
e-mail: x.yang@mdx.ac.uk

© Springer International Publishing Switzerland 2016
X.-S. Yang (ed.), *Nature-Inspired Computation in Engineering*,
Studies in Computational Intelligence 637, DOI 10.1007/978-3-319-30235-5_2

1 Introduction

Nowadays, transportation is a crucial activity for modern society, both for citizens and for business sectors. Regarding the transportation in the business world, the rapid advance of technology has made the logistic increasingly important in this area. The fact that anyone in the world can be well connected has led to transport networks that are very demanding, though such networks might be less important in the past. Today, a competitive logistic network can make a huge difference between some companies and others. On the other hand, public transport is used by almost all the population and can thus affect the quality of life. In addition, there are different kinds of public transportation systems, each one with its own characteristics. Nonetheless, all of them share the same disadvantages such as the finite capacity of the vehicles, the geographical area of coverage, and the service schedules and frequencies.

Because of their importance, the development of efficient methods for obtaining a proper logistic, or routing planning solution is a hot topic in the scientific community. In the literature, several areas of knowledge can be related to tackle with transport modelling and optimization issues. However, due to the complex nature of such transport networks, efficient methods are yet to be developed. Therefore, this work attempts to focus on artificial intelligence, one of the these active research areas.

In fact, route planning is one of most studied fields related to artificial intelligence. Problems arisen in this field are usually known as the so-called vehicle routing problems, which are a particular case of problems within combinatorial optimization. Different sorts of VRPs can lead to lots of research work annually in international conferences [1, 2], journals [3–5], technical reports [6] and edited books [7, 8].

There are many main reasons for such popularity and importance of the routing problems, however, we only highlight two reasons: the social interests they generate and their inherent scientific interests. On the one hand, routing problems are normally designed to deal with real-world situations related to the transport or logistics. Their efficient resolution can lead to profits, either social or business. On the other hand, most of the problems arising in this field have a great computational complexity, typically NP-Hard [9]. Thus, the resolution of these problems is a major challenge for researchers. Probably the most famous problems in this area are the Traveling Salesman Problem [10] and the Vehicle Routing Problem [11].

This chapter focuses on the VRP family of problems. The basic VRP consists of a set of clients, a fleet of vehicles with a limited capacity, and a known depot. The main objective of the VRP is to find the minimum number of routes with the minimum costs such that (i) each route starts and ends at the depot, (ii) each client is visited exactly by one route and (iii) the total demand of the customers visited by one route does not exceed the total capacity of the vehicle that performs the task.

Besides the basic TSP and VRP, many variations of these problems can be found in the literature. The emphasis of this work is one one of these variants: the well-known vehicle routing problem with time windows (VRPTW). In this variant, each client imposes a time window for the start of the service. The VRPTW will be described in detail in the following sections.

In line with this, several appropriate methods can be found in the literature to tackle this kind of problems in a relatively efficient way, especially for small-scale problems. Probably, the most effective techniques are the exact methods [12, 13], heuristics and metaheuristics. In our present work, our focus is nature-inspired meta-heuristics. Some classic examples of metaheuristics are tabu search [14], simulated annealing [15], ant colony optimization [16], genetic algorithms (GA) [17, 18], and particle swarm optimization [19], though most recent metaheuristic algorithms are population-based. Despite having been proposed many years ago, these techniques remain active and useful with diverse applications [20–22].

Despite of the existence of these well-known techniques, it is still necessary to design new techniques because existing methods can have some disadvantages or still struggle to cope with such tough optimization problems. In fact, new and different metaheuristic algorithms have been proposed in recent years and they have been successfully applied to a wide range of fields and problems. Some examples of these methods are the harmony search proposed by Geem et al. in 2001 [23], the cuckoo search developed by Yang and Deb in 2009 [24, 25], the firefly algorithm developed by Yang in 2008 [27] and the gravitational search algorithm presented by Rashedi in 2009 [26].

The present work focus on the the metaheuristic, called Firefly Algorithm (FA). The FA was developed by Yang in 2008 [27] as a new nature-inspired algorithm based on the flashing behaviour of fireflies. The flashing acts as a signal system to attract other fireflies. As can be shown in several surveys [28, 29], the FA has been applied in several different optimization fields and problems since its proposal, and it still attracts a lot of interests in the current scientific community [30–32]. Nevertheless, the FA has been rarely applied to any VRP problem. This lack of works, along with the growing scientific interest in bio-inspired algorithms, and the good performance shown by the FA since its proposal in 2008, has motivated its use in this study.

In addition, it is worth mentioning that several novel route optimization operators will be presented in this paper. These operators perform selective extractions of nodes in an attempt to minimize the number of routes in the current solution. For this purpose, the size of the route, the distance of the nodes from the center of gravity of the route or just random criteria are used. Specifically, the experiments presented in this chapter try to delete a route at random and then re-insert the extracted nodes on the remaining routes.

Therefore, in order to prove that our Evolutionary Discrete Firefly Algorithm (EDFA), which uses our proposed novel operators, is a promising technique to solve the well-known VRPTW, experiments with 56 instances have been conducted. In this set of experiments, the performance of several versions of the EDFA will be compared. Besides that, with the aim of drawing fair and rigorous conclusions, in addition to the conventional comparison based on the typical descriptive statistics parameters (results average, standard deviation, best result, etc.), we have also conducted two different statistical tests: the Friedman test and the Holm's test.

The rest of the chapter is organized as follows. In Sect. 2, a brief background is presented. In Sect. 3, the basic aspects of the FA are detailed. In addition, in Sect. 4, an in-depth description of the VRPTW is shown. Then, our proposed EDFA and Route optimization operators are described in Sect. 5. After that, the experimentation carried out is detailed in Sect. 6. Then, the chapter finishes with the conclusions of the study and further work in Sect. 7.

2 Related Work

As we have mentioned in the previous section, the FA is a population-based algorithm proposed in 2008 by Yang. The basic FA is based on the flashing behaviour of fireflies, and its first version was proposed for solving continuous optimization problems. Since this first implementation, the FA has been applied in a wide range of areas. Some of these areas are the continuous optimization [33, 34], multi-modal optimization [35, 36] combinatorial optimization [37], and multi-objective optimization [38].

Regarding the application fields in which the FA has proven to be effective, we can list the image processing [39], the antena design [40], civil engineering [41], robotics [42], semantic web [43], chemistry [44], and metereology [45].

In addition, several modifications and hybrid algorithms have been presented in the literature. In [46], for example, a modification called modified Firefly Algorithm is proposed. In [47, 48], on the other hand, a Chaos randomized firefly algorithm was developed. Besides that, in [49, 50] two Parallel Firefly Algorithms were presented. Regarding hybrid techniques, in [51, 52] two FAs hybridized with the GA were developed. Additionally, in [53] an ant colony hybridized with a FA was proposed. Finally, in [54] an approach was presented in which the FA was hybridized with neural networks.

In the present work, we develop a discrete version of the FA. Although the first version FA was designed for continuous problems, it has been modified many times in the literature with the intention of addressing discrete optimization problems. In [55], for instance, we can find a discrete FA adjusted to solve the class of discrete problems named Quadratic Assignment Problem. Another successful discrete FA was developed by Sayady et al. in 2010 [37] for solving minimizing the makespan for the permutation flow shop scheduling problem which is classified as a NP-Hard problem. Another discrete FA was presented in [56] by Marichelvam et al. in 2014 for the multi-objective flexible job shop scheduling problem. On the other hand, a novel evolutionary discrete FA applied to the symmetric TSP was presented in [57].

Despite the huge amount of related works, as we have pointed in the introduction, the FA has been rarely applied to any routing problem. This lack of application, along with the growing scientific interest in bio-inspired algorithms, and the good performance shown by the FA, has been the main motivation of its use in this study. Nevertheless, one of the main originalities of this work is the application field of the FA. Another novelty of our proposed approach is the use of the Hamming Distance function to measure the distance between two fireflies of the swarm. This approach has been used previously in other techniques applied to the TSP, proving its good performance [58], but it has been never used for any EDFA applied to VRPTW. In addition, the movement functions that have been used in the proposed EDFA have been never used before in the literature.

Regarding the VRPTW, the number of publications retated to this problem is really high. For this reason, we can only mention only a fraction of some recently published studies. In [59], an interesting paper published by Desaulniers et al. in 2014 can be found, in which a set of exact algorithms are presented to tackle the electric VRPTW. On other hand, Belhaiza et al. proposed in their work [60] a hybrid variable neighborhood tabu search approach for solving the VRPTW. Besides that, in 2014, a multiple ant colony system was developed for the VRPTW with uncertain travel times by Toklu et al. [61]. Finally, an interesting hybrid generational algorithm for the periodic VRPTW can be found in [62].

Finally, it is worth pointing that the set of papers and books listed in this section is only a small sample of all the related work that can be found in the literature. Because of this huge amount of related works, to summarize all the interesting papers is obviously a complex task. For this very reason, if any reader wants to extend the information presented in this work, we recommend the reading of the review paper presented in [29] about FAs. On the other hand, for additional information about the VRPTW and its solving methods, the work presented in [63, 64] is highly recommended.

3 Firefly Algorithm

The first verion of the FA was developed by Xin-She Yang in 2008 [27, 36], and it was based on the idealized behaviour of the flashing characteristics of fireflies. To understand this method in a proper way, it is important to clarify the following three idealized rules, which have been drawn from [27]:

- All the fireflies of the swarm are unisexual, and one firefly will be attracted to other ones regardless of their sex.
- Attractiveness is proportional to the brightness, which means that, for any two fire-flies, the brighter one will attract the less bright one. The attractiveness decreases as the distance between the fireflies increases. Furthermore, if one firefly is the brightest one of the swarm, it moves randomly.
- The brightness of a firefly is directly determined by the objective function of the problem under consideration. In this manner, for a maximization problem, the brightness can be proportional to the objective function value. On the other hand, for a minimization problem, it can be the reciprocal of the objective function value.

Algorithm 1: Pseudocode of the basic version of the FA.

---

1 Define the objective function $f(x)$;
2 Initialize the firefly population $X = x_1, x_2, ..., x_n$;
3 Define the light absorption coefficient γ;
4 **for** *each firefly* x_i *in the population* **do**
5 | Initialize light intensity I_i;
6 **end**
7 **repeat**
8 | **for** *each firefly* x_i *in the swarm* **do**
9 | **for** *each other firefly* x_j *in the swarm* **do**
10 | **if** $I_j > I_i$ **then**
11 | Move firelfy x_i toward x_j ;
12 | **end**
13 | Attractiveness varies with distance r via exp($-\gamma r$);
14 | Evaluate new solutions and update light intensity;
15 | **end**
16 | **end**
17 | Rank the fireflies and find the current best;
18 **until** *termination criterion reached*;
19 Rank the fireflies and return the best one;

The pseudocode of the basic version of the FA is depicted in Algorithm 1. This pseudocode was proposed by Yang in [27]. Consistent with this, there are three crucial factors to consider in the FA: the attractiveness, the distance and the movement. In the basic FA these three factors are addressed in the following way. First of all, the attractiveness of a firefly is determined by its light intensity, and it can be calculated as follows:

$$\beta(r) = \beta_0 e^{-\gamma r^2} \tag{1}$$

On the other hand, the distance r_{ij} between two fireflies i and j is determined using the Cartesian distance, and it is computed by this formula:

$$r_{ij} = ||X_i - X_j|| = \sqrt{\sum_{k=1}^{d} \left(X_{i,k} - X_{j,k} \right)^2} \tag{2}$$

where $X_{i,k}$ is the kth component of the spatial coordinate X_i of the ith firefly in the d-dimensional space. Finally, the movement of firefly i toward any other brighter firefly j is calculated as follows:

$$X_i = X_i + \beta_0 e^{-\gamma r_{ij}^2}(X_j - X_i) + \alpha(rand - 0.5) \tag{3}$$

where α is the randomization parameter and *rand* is a random number uniformly distributed in [0, 1]. On the other hand, the second term of the equation stems from the attraction assumption.

4   Vehicle Routing Problem with Time Windows

The VRPTW is an extension of the basic VRP, in which, apart from capacity constraints of each of the vehicles, each client has an associated time window $[e_i, l_i]$. This time window has a lower limit e_i and an upper limit l_i which have to be respected by all the vehicles. In other words, the service in every customer must necessarily start after e_i and l_i before. This would be the variant with hard time windows; there is also another variant that enables noncompliance with some time window (with a penzalizacin in the objective function).

Therefore, a route is not feasible if a vehicle reaches the position of any client after the upper limit of the range. By contrast, the route is feasible whether a vehicle reaches a customer before its lower limit. In this case, the client cannot be served before this limit, so that the vehicle has to wait until e_i. In addition, the central depot has also a time window, which restricts the period of activity of each vehicle in order to adapt to this range. Apart from this temporal window, it can also take into account the customer's service time. This parameter is the time that the vehicle is parked on the client while it is performing the supply. It is a factor to be taken into account to calculate if the vehicle arrives on time to the next customer.

This problem has been widely studied both in the past [63, 65, 66], and nowadays [67, 68]. One reason why the VRPTW is so interesting is its dual nature. It might be considered as a problem of two phases, one phase concerning the vehicle routing and other concerning the planning phase or customer scheduling. Another reason is its easy adaptation to the real world, because in the great majority of distribution chains, customers have strong temporal constraints that have to be fulfilled. For example, in the distribution of the press or of perishable foods these windows are really necessary.

Regarding the mathematical formulation of VRPTW, it can take several forms, using more or less variables [69, 70]. One of the most interesting formulations can be found in [71].

5   Our Proposed Approach for Solving the VRPTW

In this section, the description of our EDFA for the VRPTW is provided (Sect. 5.1). Besides that, a detailed description of the proposed novel route optimization operators can be found in Sect. 5.2.

5.1   An Evolutionary Discrete Firefly Algorithm

It is worth mentioning that the original FA was primarily developed for solving continuous optimization problems. This is the reason because the classic FA cannot be applied directly to solve any discrete problem, such as the VRPTW. Hence, some modifications in the flowchart of the basic FA must be conducted with the aim of preparing it for tackling the VRPTW.

First of all, in the proposed EDFA, each firefly in the swarm represents a possible and feasible solution for the VRPTW. In addition, as it is well-known, the VRPTW is a minimization problem. For this reason, the most attractive fireflies are those with a lower objective function value. The concept of light absorption is also represented in this version of the FA. In this case, $\gamma = 0.95$, and this parameter is used in the same way as has been depicted in Eq. (3). This parameter has been set following the guidelines proposed in several studies of the literature [27, 36].

Furthermore, as has been mentioned in the introduction of this paper, the distance between two fireflies is calculated using the well-known Hamming distance. The Hamming distance between two fireflies is the number of non-corresponding elements in the sequence. In the experimentation, VRPTW solutions are represented by a giant-tour, which consists of the client identifiers, being 0 the depot. Thus, the Hamming distance is calculated from the comparison of the order who have the clients in the giant-tour (excluding the depot). For example, given two solutions (or firefly) problem consisting of 7 nodes:

$$x_1 : \{0, 1, 2, 5, 0, 3, 4, 6, 7, 0\} \rightarrow 1, 2, \mathbf{5}, \mathbf{3}, 4, \mathbf{6}, \mathbf{7},$$

$$x_2 : \{0, 1, 2, 6, 0, 7, 4, 0, 5, 3, 0\} \rightarrow 1, 2, \mathbf{6}, \mathbf{7}, 4, \mathbf{5}, \mathbf{3},$$

the Hamming Distance between x_1 and x_2 would be 4.

This same example serves to analyze the brightness (light intensity I_i) of a firefly. In this case, the fitness function used is traditional for the VRP. It has two hierarchical objectives: first the number of routes and as a secondary objective the total traveled distance. As shown in the above example, firefly x_1 is better than the firefly x_2 because the former has fewer routes than the second; 2 versus 3. Thus x_2 will be attracted to x_1 using the proposed route optimization operator.

Finally, the movement of firefly i attracted to another brighter firefly j is determined by

$$n = \text{Random}\left(2, r_{ij} \cdot \gamma^g\right) \tag{4}$$

where r_{ij} is the Hamming Distance between firefly i and firefly j, and g is the iteration number. In this case, the length of the movement of a firefly will be a random number between 2 and $r_{ij} \cdot \gamma^g$. This value is used to generate n successors from the solution corresponding to the firefly to be moved. Once all successors are generated, the best of them is selected to replace the original firefly. For comparison of different alternatives, two criteria for selecting the best successor will be used: the successor with the best objective function value, or the successor with the lower Hamming distance towards the firefly j (the one that is used as reference to perform the movement of firefly i).

In the proposed EDFA, a single operator to simulate the movement of fireflies is used. This operator is based on the description given in Sect. 5.2 with the following features:

- The *ejectionPool* is initialized with all the nodes assigned to a randomly selected route.
- To speed up the process, the optimization of the remaining routes and the re-insertion into the nearest route phases are not performed.

Furthermore, Regarding the termination criterion, each technique finishes its execution when it reach the generation (iteration) 101, or when there are 20 generations without any improvement in the best solution found.

Finally, after conducting an empirical analysis, the "first-movement" criterion is used to stop the process of attracting a firefly in each global iteration. In this sense, when a firefly at x_i is attracted by other firefly x_j, the movement of x_i during the current iteration is finished. After that, the algorithm continues with the process of the firefly x_{i+1}. This scheme accelerates the whole process without significantly affecting the quality of the final solution obtained by the algorithm.

5.2   Description of the Proposed Operators

In the context of VRP and its variants there are a number of operators (Or-opt, 2-opt, String-reallocation, String-exchange, GENI-exchange, GENI-CROSS, etc. see [72]) whose objective focuses on the improvement of routes through the exchange of nodes (clients) or paths (sequences of clients) both for a single route and between small groups of routes. These operators perform small modifications to the current solution which allow to control the algorithm computational complexity and runtime. While processing time is an important element, these operators focus their analysis on solutions close to the current solution (intensification capacity) by limiting the space of solutions that are able to explore. This may limit the exploration of the search space avoiding the movement to areas that might contain more promising solutions (diversification ability). In addition, these operators have a limited or negligible capacity to reduce the number of routes since only on rare occasions, the movement of a set of nodes between two routes can leave one of them empty, allowing reduction of the number of routes in the current solution.

On the other hand, there are heuristics that focus their efforts on minimizing the number of routes. These techniques, which have their origin in the "ejection chains" method [73], carried out processes of extraction and reinsertion of nodes on the routes of the current solution. This methods could also remove a complete route in order to minimize the number of routes. In the latter case, probably one of the most representative and successful route minimization heuristics was developed by Nagata and Brysy [74].

Taking as inspiration the concept of "ejection chains", a family of operators whose objective is the reduction of the number of routes has been presented in this work. These operators combine the "ejection chains" technique with other simple measures (such as the size of a route and the proximity with respect to the "centre of gravity of a route"). The proposed operators are initially designed to be integrated into local search processes. In this way, the developed operators increase the diversification

---

Algorithm 2: Pseudocode of the route minimization operator.

input : $Solution_{current}$, $optimizeRoutes$, $proximityReinsertion$
1 $ejectionPool = initEjectionPool(Solution_{current})$;
2 $Solution_{new} = removeEmptyRoutes(Solution_{current})$;
3 **if** $optimizeRoutes$ **then**
4 $\quad|\quad$ $optimizeRoutes(Solution_{new})$;
5 **end**
6 **if** $proximityReinsertion$ **then**
7 $\quad|\quad$ $reinsert(ejectionPool,Solution_{new})$;
8 **end**
9 **if** $ejectionPool \neq \oslash$ **then**
10 $\quad|\quad$ $Solution_{new} = parallelReconstruction(ejectionPool,Solution_{new})$;
11 **end**
12 **if** $Solution_{new}$ better than $Solution_{current}$ **then**
13 $\quad|\quad$ $Solution_{current} = Solution_{new}$;
14 **end**
   output: $Solution_{current}$

---

ability of the traditional node and arc interchange based operators. After describing the basic notion of the proposed operators, the main characteristics of them are depicted. The descriptions of the operators focus on VRPTW, but they could be easily adapted for any other variant of the VRP.

VRPTW construction heuristics focus their efforts on the generation of an initial solution in a fast and efficient way. This fact hinders the ability to explore the space of solutions, taking irreversible decisions when assigning clients to a vehicle, and sort them in a route. For that reason, after applying a construction heuristic, improvement processes are needed. These processes review allocation and sort decisions to obtain better solutions. This argument is consistent due to the nature of the VRP, but could even explicitly confirmed if the construction process could analyze in detail the structure of the generated routes and the location of customers. For example, after obtaining a solution to a VRP, a person might suggest changes (in the allocations made) visually analyzing the solution. This process could be based on simple calculations to analyze the number of clients of a route, or the proximity between customers that form a route. This notion, combined with random behavior, has been used as a basis for designing the new operators for VRPTW.

The description of the proposed operators is shown in Algorithm 2.

- First the *ejectionPool* is initialized (line 1). This structure is composed by the nodes who are extracted from their original route and will be reinserted again to create the new solution. The construction of the *ejectionPool* allows several variants to extract nodes from its original location, for instance:
  - Extract the nodes further away from the center of gravity of their original route.
  - Extract all the nodes belonging to the smallest route(s).
  - Extract all the nodes from a randomly selected route.

In the work presented here only the third variant is applied, but other variants can be defined.

- Once the *ejectionPool* is initialized "empty routes elimination" step is performed (line 3). In this phase the routes that are empty after node extraction are eliminated.
- After the removal of empty routes, an optional optimization process is done (line 4). This process is based on an intra-route operator (it modifies a single route). Its aim is to increase the chances for reinserting the extracted nodes. For example the use of Or-opt or 2-opt is suggested. In the experimentation conducted in this work this step is skipped to speed up the overall process.
- After optimizing the remaining routes, the algorithm continues with the "reinsertion phase" (line 7). This phase is also optional. The basic idea would be to reinsert each of the nodes that are part of the *ejectionPool* in its "nearest" route. To perform this reinsertion in an efficient way, the use of neighbor lists is recommended [75].
- As a last step, the final reconstruction of the new solution is performed (line 10). In this phase, a parallel construction heuristic is used. This heuristics combines the routes of the current solution and nodes remaining in the "ejection pool". After invoking the parallel construction heuristic all the nodes are again assigned to a route and the process ends returning the new solution. The reconstruction algorithm could be any construction technique but in this case the one proposed by Campbell and Savelsbergh [76] is used.

With the scheme described above, it can be seen that this new type of operator performs a more complex process than traditional node and arc exchange operators. This can affect the runtime but the proposed operator possesses a good ability to reduce the number of routes that are in the current solution.

Normally, the process of minimizing the number of routes is the last step of a heuristic or a metaheuristic. Actually, in some cases it is run as a completely separate process. But with the new proposed operators this process can be integrated implicitly in the optimization algorithm. In fact, the new operator can be a perfect complement to increase the diversification ability in the population. The proposed operator will be used in the proposed EDFA algorithm to implement the movement of the fireflies in the swarm.

6 Experiments and Results

In order to test our proposed approaches properly, the 100 customers Solomon's problems and instances will be used [77]. This set of problems consists of 56 instances classified into 6 categories (C1, C2, R1, R2, RC1 y RC2) which differ in the geographical distributions of the customers, the capacities of the vehicles and the compatibility of the time windows.

Table 1 Results of $EDFA_{FR-HD}$

Class	T	AVG_V	SD_V	AVG_D	SD_D
C1	3792	11.022	0.050	1716.296	51.919
C2	5119	3.975	0.056	1099.617	25.176
R1	4339	14.033	0.045	1567.214	16.077
R2	7608	3.182	0.000	1325.060	18.094
RC1	2672	14.225	0.105	1847.529	13.604
RC2	4910	3.800	0.112	1600.324	21.278

Although there are VRPTW benchmarks with larger problems instances (such as Gehring & Homberger's[1]), the objective of the work presented focuses on analyzing the adaptation of the EDFA algorithm to the VRPTW. For this reason, Solomon's benchmark is adequate and representative to analyze the behavior of the EDFA applied to VRPTW.

All the tests conducted in this work have been performed on an Intel Core i5 2410 laptop, with 2.30 GHz and a RAM of 4 GB. Java has been used as the programming language.

It is important to point out that the objective function used for the VRPTW is the classic one, which prioritizes the minimization of the routes number, leaving the traveled distance as the second optimizing criterion.

The experimentation has been performed with 4 variants of the proposed EDFA described in Sect. 5.2: $EDFA_{FR-OF}$, $EDFA_{HR-OF}$, $EDFA_{FR-HD}$ and $EDFA_{HR-HD}$. Such variants differ in the use of two criteria for initializing the swarn of fireflies (Full Random = 100 % random and Half Random = 50 % random +50 % good solutions) and two criteria to select the best successor to move a firefly x_i towards a firefly x_j (Objective Function = successor with the best value of the objective function, Hamming Distance = successor with the lower Hamming distance from x_j). To create the good initial solutions, Solomon's I1 construction heuristic [77] has been used. Additionally, the initial population size has been set to 50 fireflies. Finally, all the variants of the EDFA have been executed 20 times.

The results of the experimentation are shown in Tables 1, 2, 3 and 4. All the tables have the same structure: one row for each class of the Solomon's bechmark (summarizing the results of all the instances of a class) and five columns. Each column corresponds to the average runtime for all the instances of each class (T, in seconds), and average (AVG) and standard deviation (SD) for the number of routes (V) and the total cumulative distance (D).

Table 1 presents the results obtained by $EDFA_{FR-HD}$. This version of the algorithm is characterized by using a completely random initial population. Best successor of firefly x_i is chosen based on the Hamming distance between the successor

---

[1] https://www.sintef.no/projectweb/top/vrptw/homberger-benchmark/.

Table 2 Results of $EDFA_{FR-OF}$

Class	T	AVG_V	SD_V	AVG_D	SD_D
C1	3089	10.689	0.093	1513.885	26.913
C2	4300	3.900	0.105	1031.001	32.767
R1	3605	13.667	0.084	1506.030	5.671
R2	6629	3.218	0.050	1288.694	8.178
RC1	2166	13.650	0.105	1734.507	21.553
RC2	4364	3.750	0.088	1590.146	14.793

Table 3 Results of $EDFA_{HR-HD}$

Class	T	AVG_V	SD_V	AVG_D	SD_D
C1	2887	10.000	0.000	914.323	2.568
C2	4634	3.000	0.000	671.709	2.771
R1	4146	13.250	0.059	1464.889	10.628
R2	7387	3.182	0.000	1261.209	2.617
RC1	2384	12.975	0.105	1633.282	16.704
RC2	4914	3.500	0.000	1499.629	1.925

Table 4 Results of $EDFA_{HR-OF}$

Class	T	AVG_V	SD_V	AVG_D	SD_D
C1	2480	10.000	0.000	907.105	0.615
C2	3978	3.000	0.000	666.225	2.360
R1	3341	13.188	0.042	1442.712	3.956
R2	6693	3.182	0.000	1243.179	2.507
RC1	2013	12.969	0.063	1568.936	6.000
RC2	4407	3.500	0.000	1490.360	4.891

and firefly x_j (which attracts x_i). According to the experimentation conducted, this variant is the one that gets worse results both in number of vehicles and traveled distance. This confirms the importance of the quality of the initial solution in the VRPTW. On the other hand, the results also serve to justify the Hamming distance offers a worse performance than the objective function, to choose the best successor to move a firefly. Like the other variants, standard deviation in relation to the number of vehicles is not very high.

In Table 2 the results of $EDFA_{FR-OF}$ are presented. In this case the initial population has been generated 100 % at random. Regarding the selection of successors this variant uses the same criteria as $EDFA_{HR-OF}$. This variant improves the foregoing initialization issue and its results are slightly better. However, the two variants are tied for the number of vehicles in the R2 class.

$EDFA_{FR-HD}$ results are shown in Table 3. In this case the initial population is generated 50 % randomly and 50 % using Solomon's I1 construction heuristic. The best successor to every movement of a firefly is chosen based on Hamming distance. As we can see, this variant performs better than the previous two in terms of the number of vehicles and distance traveled. The improvement is mainly due to the quality of the initial solutions. This confirms the relevance of the initial solution in the VRPTW. Furthermore, by analyzing standard deviations, it can be seen that the values are lower. This implies that this method is also more robust.

Finally, Table 4 shows the results of $EDFA_{HR-OF}$. This variant of the EDFA combines the initialization process of $EDFA_{HR-XX}$ with the selection of the best successor used by $EDFA_{XX-OF}$. This variant is the one that gets the best results. Always gets the best results in terms of traveled distance, and ties with $EDFA_{HR-HD}$ in terms of the number of vehicles (except for the R1 class). Given the characteristics of this variant (and always according to the experimentation carried), it confirms that the objective function is better than the Hamming distance in selecting successors. Furthermore, the quality of the initial solution also affects the final solution: the better the quality of the initial solution, the better the final solution.

To summarize, Table 5 shows the comparison of all variants and the difference from the $EDFA_{HR-OF}$ (which reported the best results). From the table, it can be observed that $EDFA_{HR-OF}$ obtains the best results in terms of distance, having a draw with $EDFA_{HR-HD}$ regarding the number of vehicles. To finish this preliminary analysis, in relation to execution times, all values are quite similar. They are between 300 and 600 s to solve an instance of a problem.

Once the results of the experimentation have been presented, two statistical tests (using the number of vehicles and traveled distance) have been made. These tests are based on the guidelines suggested by Derrac et al. [78]. The objective of this task is to ensure that comparisons between the different variants of the EDFA are fair and objective. First, the non-parametric Friedmans test for multiple comparison was conducted. This test aims to check for significant differences between the four variants of the EDFA.

Table 6 shows the average ranking obtained for each variant (the lower the value, the better the performance of the variant). The test has been conducted for both criteria of the objective function: the number of vehicles and total traveled distance. Regarding the number of vehicles, the resulting Friedman statistic has been 9.95. Taking into account that the confidence interval has been stated at the 97.5 % confidence level, the critical point in a χ^2 distribution with 3 degrees of freedom is 9.348. Since $9.95 > 9.348$, it can be concluded that there are significant differences among the results reported by the four compared algorithms, being $EDFA_{HR-OF}$ the one with the lowest rank. Finally, for this Friedman test, the computed p-value has been 0.018996.

Once discovered significant differences in the number of vehicles, it is appropriate to compare technique by technique. For that reason, a post-hoc Holm's test, using $EDFA_{HR-OF}$ as reference (which ranks first in the number of vehicles), has been made. The results of this test are shown in Table 7. As can be seen, only for $EDFA_{FR-OF}$ adjusted and unadjusted p-values are simultaneously less than or equal

Table 5 Summary of the results and comparison with $EDFA_{HR-OF}$

	$EDFA_{HR-OF}$				$EDFA_{HR-HD}$				$EDFA_{FR-OF}$				$EDFA_{FR-HD}$			
	AVG_V	$\%_V$	AVG_D	$\%_D$	AVG_V	$\%_V$	AVG_D	$\%_D$	AVG_V	$\%_V$	AVG_D	$\%_D$	AVG_V	$\%_V$	AVG_D	$\%_D$
C1	**10.000**	0.000	**907.105**	0.000	10.000	0.000	914.323	0.008	10.689	0.064	1513.885	0.401	11.022	0.093	1716.296	0.471
C2	**3.000**	0.000	**666.225**	0.000	3.000	0.000	671.709	0.008	3.900	0.231	1031.001	0.354	3.975	0.245	1099.617	0.394
R1	**13.188**	0.000	**1442.712**	0.000	13.250	0.005	1464.889	0.015	13.667	0.035	1506.030	0.042	14.033	0.060	1567.214	0.079
R2	**3.182**	0.000	**1243.179**	0.000	3.182	0.000	1261.209	0.014	3.218	0.011	1288.694	0.035	3.182	0.000	1325.060	0.062
RC1	**12.969**	0.000	**1568.936**	0.000	12.975	0.000	1633.282	0.039	13.650	0.050	1734.507	0.095	14.225	0.088	1847.529	0.151
RC2	**3.500**	0.000	**1490.360**	0.000	3.500	0.000	1499.629	0.006	3.750	0.067	1590.146	0.063	3.800	0.079	1600.324	0.069

Table 6 Average ranking obtained by the Friedman's test

Algorithm	AVG_V	AVG_D
$EDFA_{HR-OF}$	1.500	1.166
$EDFA_{HR-HD}$	1.916	1.833
$EDFA_{FR-OF}$	3.000	3.000
$EDFA_{FR-HD}$	3.583	4.000

Table 7 Adjusted and unadjusted p-values of Holm's test for the number of vehicles

Algorithm	Adjusted p	Unadjusted p
$EDFA_{FR-OF}$	0.005189	0.015566
$EDFA_{FR-HD}$	0.044171	0.088343
$EDFA_{HR-HD}$	0.576150	0.576150

Table 8 Adjusted and unadjusted p-values of Holm's test for the total traveled distance

Algorithm	Adjusted p	Unadjusted p
$EDFA_{FR-OF}$	0.000144	0.000432
$EDFA_{FR-HD}$	0.013906	0.027813
$EDFA_{HR-HD}$	0.371093	0.371093

to 0.05. Therefore, it can be confirmed statistically that the differences in the number of routes for all variants regarding $EDFA_{HR-OF}$ are only significant for $EDFA_{HR-OF}$.

The statistical test of the number of vehicles show no significant differences. For that reason, a new statistical analysis has been performed. This second analysis has been focused on the total traveled distance. This has involved the implementation of new Friedman's and Holm's tests (Table 8).

For the traveled distance, the resulting Friedman statistic has been 17. In this case $17 > 9.348$, so it can be concluded that there are also significant differences among the results reported by the compared algorithms, being $EDFA_{HR-OF}$ the one with the lowest rank. Finally, regarding this Friedman test, the computed p-value has been 0.000707. In addition, a new Holm test using traveled distance has been performed. $EDFA_{HR-OF}$ has been the reference again. The results confirm the existence of significant differences with respect to $EDFA_{FR-HD}$ and $EDFA_{FR-OF}$; but no significant differences regarding $EDFA_{HR-HD}$ exists. These results confirm the superiority of the initialization good solutions with respect to 100 % random initialization. Finally, after combining the results of the two rankings (number of routes and total distance), it may conclude that the $EDFA_{HR-OF}$ variant is the one that gets the best results for the experimentation conducted.

7 Conclusions

In this work, an Evolutionary Discrete Firefly Algorithm applied to the well-known Vehicle Routing Problem with Time Windows has been presented. The proposed technique presents some novelties, such as the use of the Hamming Distance to measure the distance between two different fireflies. Another interesting originality is the novel route optimization operators that have been developed for the EDFA. These operators perform selective extractions of nodes in an attempt to minimize the number of routes in the actual solution. For this, the size of the route, the distance of the nodes from the center of gravity of the route or just random criteria are used. Specifically, the experimentation conducted in the work presented uses an operator that removes a random selected route and then try to reinsert the extracted nodes in the remaining routes.

In order to demonstrate that the proposed EDFA and the developed route optimization operators are promising approaches, the perfomance of the presented EDFA has been compared with those obtained by several versions of the EDFA. For this comparison the 56 instances of 100 customers Solomon's VRPTW bechmark have been used. Besides that, in order to obtain fair conclusions, two different statistical tests have been performed: the Friedman's Test and the Holm's Test.

As for future work, we plan to extend the experimentation of this study, comparing the performance of the proposed EDFA with those presented by some recently proposed metaheuristic, such as the Bat Algorithm, or the Golden Ball Algorithm [79]. In addition, we intend to use the novel route optimization operators proposed in this work in other recent and classic techniques, such as the Simulated Annealing, or Genetic Algorithm.

Acknowledgments This project was supported by the European Unions Horizon 2020 research and innovation programme through the TIMON: Enhanced real time services for optimized multimodal mobility relying on cooperative networks and open data project (636220); as well as by the projects TEC2013-45585-C2-2-R from the Spanish Ministry of Economy and Competitiveness, and PC2013-71A from the Basque Government.

References

1. Soonpracha, K., Mungwattana, A., Manisri, T.: A re-constructed meta-heuristic algorithm for robust fleet size and mix vehicle routing problem with time windows under uncertain demands. In: Proceedings of the 18th Asia Pacific Symposium on Intelligent and Evolutionary Systems, pp. 347–361, Springer (2015)
2. Wen, Z., Dong, X., Han, S.: An iterated local search for the split delivery vehicle routing problem. In: International Conference on Computer Information Systems and Industrial Applications, Atlantis Press (2015)
3. Escobar, J.W., Linfati, R., Toth, P., Baldoquin, M.G.: A hybrid granular tabu search algorithm for the multi-depot vehicle routing problem. J. Heuristics **20**(5), 483–509 (2014)
4. Lin, C., Choy, K.L., Ho, G.T., Chung, S., Lam, H.: Survey of green vehicle routing problem: past and future trends. Expert Syst. Appl. **41**(4), 1118–1138 (2014)

5. Reed, M., Yiannakou, A., Evering, R.: An ant colony algorithm for the multi-compartment vehicle routing problem. Appl. Soft Comput. **15**, 169–176 (2014)
6. Coelho, L.C., Renaud, J., Laporte, G.: Road-based goods transportation: a survey of real-world applications from 2000 to 2015. Technical report, Technical Report FSA-2015-007, Québec, Canada (2015)
7. Toth, P., Vigo, D.: The vehicle routing problem. Soc. Ind. Appl. Math. (2015)
8. Laporte, G., Ropke, S., Vidal, T.: Heuristics for the vehicle routing problem. Veh. Routing Prob. Methods Appl. **18**, 87 (2014)
9. Lenstra, J.K., Kan, A.: Complexity of vehicle routing and scheduling problems. Networks **11**(2), 221–227 (1981)
10. Lawler, E.L.: The traveling salesman problem: a guided tour of combinatorial optimization. Wiley-interscience series in discrete mathematics (1985)
11. Dantzig, G.B., Ramser, J.H.: The truck dispatching problem. Manage. Sci. **6**(1), 80–91 (1959)
12. Laporte, G.: The traveling salesman problem: an overview of exact and approximate algorithms. Eur. J. Oper. Res. **59**(2), 231–247 (1992)
13. Laporte, G.: The vehicle routing problem: an overview of exact and approximate algorithms. Eur. J. Oper. Res. **59**(3), 345–358 (1992)
14. Glover, F.: Tabu search, part i. ORSA J. Comput. **1**(3), 190–206 (1989)
15. Kirkpatrick, S., Gellat, C., Vecchi, M.: Optimization by simmulated annealing. Science **220**(4598), 671–680 (1983)
16. Dorigo, M., Blum, C.: Ant colony optimization theory: a survey. Theoret. Comput. Sci. **344**(2), 243–278 (2005)
17. Goldberg, D.: Genetic algorithms in search, optimization, and machine learning. Addison-Wesley Professional (1989)
18. De Jong, K.: Analysis of the behavior of a class of genetic adaptive systems. Ph.D. thesis, University of Michigan, Michigan, USA (1975)
19. Kennedy, J., Eberhart, R., et al.: Particle swarm optimization. In: Proceedings of IEEE International Conference on Neural Networks, vol. 4, pp. 1942–1948, Perth, Australia (1995)
20. Rodriguez, A., Gutierrez, A., Rivera, L., Ramirez, L.: Rwa: Comparison of genetic algorithms and simulated annealing in dynamic traffic. In: Advanced Computer and Communication Engineering Technology, pp. 3–14, Springer (2015)
21. Cao, B., Glover, F., Rego, C.: A tabu search algorithm for cohesive clustering problems. J. Heuristics 1–21 (2015)
22. İnkaya, T., Kayalıgil, S., Özdemirel, N.E.: Ant colony optimization based clustering methodology. Appl. Soft Comput. **28**, 301–311 (2015)
23. Geem, Z.W., Kim, J.H., Loganathan, G.: A new heuristic optimization algorithm: harmony search. Simulation **76**(2), 60–68 (2001)
24. Yang, X.S., Deb, S.: Cuckoo search via lévy flights. In: IEEE World Congress on Nature & Biologically Inspired Computing, pp. 210–214 (2009)
25. Yang, X.S., Deb, S.: Engineering optimisation by cuckoo search. Int. J. Math. Model. Numer. Optim. **1**(4), 330–343 (2010)
26. Rashedi, E., Nezamabadi-Pour, H., Saryazdi, S.: Gsa: a gravitational search algorithm. Inf. Sci. **179**(13), 2232–2248 (2009)
27. Yang, X.S.: Nature-inspired metaheuristic algorithms. Luniver press, Bristol (2008)
28. Fister, I., Yang, X.S., Fister, D., Fister Jr, I.: Firefly algorithm: a brief review of the expanding literature. In: Cuckoo Search and Firefly Algorithm, pp. 347–360, Springer (2014)
29. Fister, I., Fister Jr, I., Yang, X.S., Brest, J.: A comprehensive review of firefly algorithms. Swarm Evol. Comput. (2013)
30. Ma, Y., Zhao, Y., Wu, L., He, Y., Yang, X.S.: Navigability analysis of magnetic map with projecting pursuit-based selection method by using firefly algorithm. Neurocomputing (2015)
31. Liang, R.H., Wang, J.C., Chen, Y.T., Tseng, W.T.: An enhanced firefly algorithm to multi-objective optimal active/reactive power dispatch with uncertainties consideration. Int. J. Electr. Power Energy Syst. **64**, 1088–1097 (2015)

32. Zouache, D., Nouioua, F., Moussaoui, A.: Quantum-inspired firefly algorithm with particle swarm optimization for discrete optimization problems. Soft Comput. 1–19 (2015)
33. Yang, X.S.: Metaheuristic optimization: algorithm analysis and open problems. In: Experimental Algorithms, pp. 21–32, Springer (2011)
34. Yang, X.S.: Efficiency analysis of swarm intelligence and randomization techniques. J. Comput. Theoret. Nanosci. 9(2), 189–198 (2012)
35. Das, S., Maity, S., Qu, B.Y., Suganthan, P.N.: Real-parameter evolutionary multimodal optimizationa survey of the state-of-the-art. Swarm Evol. Comput. 1(2), 71–88 (2011)
36. Yang, X.S.: Firefly algorithms for multimodal optimization. In: Stochastic algorithms: foundations and applications, pp. 169–178, Springer (2009)
37. Sayadi, M., Ramezanian, R., Ghaffari-Nasab, N.: A discrete firefly meta-heuristic with local search for makespan minimization in permutation flow shop scheduling problems. Int. J. Ind. Eng. Comput. 1(1), 1–10 (2010)
38. Abedinia, O., Amjady, N., Naderi, M.S.: Multi-objective environmental/economic dispatch using firefly technique. In: IEEE International Conference on Environment and Electrical Engineering, pp. 461–466 (2012)
39. Zhang, Y., Wu, L.: A novel method for rigid image registration based on firefly algorithm. Int. J. Res. Rev. Soft Intell. Comput. (IJRRSIC) 2(2), 141–146 (2012)
40. Basu, B., Mahanti, G.K.: Fire fly and artificial bees colony algorithm for synthesis of scanned and broadside linear array antenna. Prog. Electromagnet. Res. B 32, 169–190 (2011)
41. Talatahari, S., Gandomi, A.H., Yun, G.J.: Optimum design of tower structures using firefly algorithm. Struct. Des. Tall Spec. Buildings 23(5), 350–361 (2014)
42. Jakimovski, B., Meyer, B., Maehle, E.: Firefly flashing synchronization as inspiration for self-synchronization of walking robot gait patterns using a decentralized robot control architecture. In: Architecture of Computing Systems-ARCS 2010, pp. 61–72, Springer (2010)
43. Pop, C.B., Rozina Chifu, V., Salomie, I., Baico, R.B., Dinsoreanu, M., Copil, G.: A hybrid firefly-inspired approach for optimal semantic web service composition. Scalable Comput. Pract. Exp. 12(3), 363–370 (2011)
44. Fateen, S.E.K., Bonilla-Petriciolet, A., Rangaiah, G.P.: Evaluation of covariance matrix adaptation evolution strategy, shuffled complex evolution and firefly algorithms for phase stability, phase equilibrium and chemical equilibrium problems. Chem. Eng. Res. Des. 90(12), 2051–2071 (2012)
45. Santos, A.F., Campos Velho, H.F., Luz, E.F., Freitas, S.R., Grell, G., Gan, M.A.: Firefly optimization to determine the precipitation field on south america. Inverse Prob. Sci. Eng. 21(3), 451–466 (2013)
46. Tilahun, S.L., Ong, H.C.: Modified firefly algorithm. J. Appl. Math. 2012, 1–12 (2012)
47. Gandomi, A., Yang, X.S., Talatahari, S., Alavi, A.: Firefly algorithm with chaos. Commun. Nonlinear Sci. Numer. Simul. 18(1), 89–98 (2013)
48. Coelho, L.D.S., de Andrade Bernert, D.L., Mariani, V.C.: A chaotic firefly algorithm applied to reliability-redundancy optimization. In: IEEE Congress on Evolutionary Computation, IEEE, pp. 517–521 (2011)
49. Subutic, M., Tuba, M., Stanarevic, N.: Parallelization of the firefly algorithm for unconstrained optimization problems. Latest Adv. Inf. Sci. Appl. 22, 264–269 (2012)
50. Husselmann, A.V., Hawick, K.: Parallel parametric optimisation with firefly algorithms on graphical processing units. In: Proceedings International Conference on Genetic and Evolutionary Methods, pp. 77–83 (2012)
51. Farahani, S.M., Abshouri, A.A., Nasiri, B., Meybodi, M.: Some hybrid models to improve firefly algorithm performance. Int. J. Artif. Intell. 8(S12), 97–117 (2012)
52. Luthra, J., Pal, S.K.: A hybrid firefly algorithm using genetic operators for the cryptanalysis of a monoalphabetic substitution cipher. In: IEEE World Congress on Information and Communication Technologies, pp. 202–206 (2011)
53. Aruchamy, R., Vasantha, K.: A comparative performance study on hybrid swarm model for micro array data. Int. J. Comput. Appl. 30(6), 10–14 (2011)

54. Hassanzadeh, T., Faez, K., Seyfi, G.: A speech recognition system based on structure equivalent fuzzy neural network trained by firefly algorithm. In: IEEE International Conference on Biomedical Engineering, pp. 63–67 (2012)
55. Durkota, K.: Implementation of a discrete firefly algorithm for the qap problem within the sage framework. BSc thesis, Czech Technical University (2011)
56. Marichelvam, M.K., Prabaharan, T., Yang, X.S.: A discrete firefly algorithm for the multi-objective hybrid flowshop scheduling problems. EEE Trans. Evol. Comput. **18**(2), 301–305 (2014)
57. Jati, G.K., et al.: Evolutionary discrete firefly algorithm for travelling salesman problem. In: Adaptive and Intelligent Systems (2011)
58. Zhou, L., Ding, L., Qiang, X.: A multi-population discrete firefly algorithm to solve tsp. In: Bio-Inspired Computing-Theories and Applications, pp. 648–653, Springer (2014)
59. Desaulniers, G., Errico, F., Irnich, S., Schneider, M.: Exact algorithms for electric vehicle-routing problems with time windows. Les Cahiers du GERAD G-2014-110, GERAD, Montréal, Canada (2014)
60. Belhaiza, S., Hansen, P., Laporte, G.: A hybrid variable neighborhood tabu search heuristic for the vehicle routing problem with multiple time windows. Comput. Oper. Res. **52**, 269–281 (2014)
61. Toklu, N.E., Gambardella, L.M., Montemanni, R.: A multiple ant colony system for a vehicle routing problem with time windows and uncertain travel times. J. Traffic Logist. Eng. **2**(1), 5–8 (2014)
62. Nguyen, P.K., Crainic, T.G., Toulouse, M.: A hybrid generational genetic algorithm for the periodic vehicle routing problem with time windows. J. Heuristics **20**(4), 383–416 (2014)
63. Kallehauge, B., Larsen, J., Madsen, O.B., Solomon, M.M.: Vehicle Routing Problem with Time Windows. Springer, New York (2005)
64. Gendreau, M., Tarantilis, C.D.: Solving large-scale vehicle routing problems with time windows: The state-of-the-art, Cirrelt (2010)
65. Potvin, J.Y., Bengio, S.: The vehicle routing problem with time windows part ii: genetic search. INFORMS J. Comput. **8**(2), 165–172 (1996)
66. Bräysy, O., Gendreau, M.: Vehicle routing problem with time windows, part i: route construction and local search algorithms. Transp. Sci. **39**(1), 104–118 (2005)
67. Afifi, S., Guibadj, R.N., Moukrim, A.: New lower bounds on the number of vehicles for the vehicle routing problem with time windows. In: Integration of AI and OR Techniques in Constraint Programming, pp. 422–437, Springer (2014)
68. Agra, A., Christiansen, M., Figueiredo, R., Hvattum, L.M., Poss, M., Requejo, C.: The robust vehicle routing problem with time windows. Comput. Oper. Res. **40**(3), 856–866 (2013)
69. Azi, N., Gendreau, M., Potvin, J.Y.: An exact algorithm for a single-vehicle routing problem with time windows and multiple routes. Eur. J. Oper. Res. **178**(3), 755–766 (2007)
70. Bräysy, O., Gendreau, M.: Tabu search heuristics for the vehicle routing problem with time windows. Top **10**(2), 211–237 (2002)
71. Cordeau, J.F., Desaulniers, G., Desrosiers, J., Solomon, M.M., Soumis, F.: Vrp with time windows. Veh. Routing Prob. **9**, 157–193 (2001)
72. Bräysy, O., Gendreau, M.: Vehicle routing problem with time windows, part I: route construction and local search algorithms. Transp. Sci. **39**(1), 104–118 (2005)
73. Rego, C.: Node-ejection chains for the vehicle routing problem: sequential and parallel algorithms. Parallel Comput. **27**(3), 201–222 (2001)
74. Nagata, Y., Brysy, O.: A powerful route minimization heuristic for the vehicle routing problem with time windows. Oper. Res. Lett. **37**(5), 333–338 (2009)
75. Irnich, S.: A unified modeling and solution framework for vehicle routing and local search-based metaheuristics. INFORMS J. Comput. **20**(2), 270–287 (2008)
76. Campbell, A.M., Savelsbergh, M.: Efficient insertion Heuristics for vehicle routing and scheduling problems. Transp. Sci. **38**(3), 369–378 (2004)
77. Solomon, M.M.: Algorithms for the vehicle routing and scheduling problems with time window constraints. Oper. Res. **35**(2), 254–265 (1987)

78. Derrac, J., García, S., Molina, D., Herrera, F.: A practical tutorial on the use of nonpara-metric statistical tests as a methodology for comparing evolutionary and swarm intelligence algorithms. Swarm Evol. Comput. **1**(1), 3–18 (2011)
79. Osaba, E., Diaz, F., Onieva, E.: Golden ball: a novel meta-heuristic to solve combinatorial optimization problems based on soccer concepts. Appl. Intell. **41**(1), 145–166 (2014)

The Plant Propagation Algorithm for Discrete Optimisation: The Case of the Travelling Salesman Problem

Birsen İ. Selamoğlu and Abdellah Salhi

Abstract The Plant Propagation algorithm (PPA), has been demonstrated to work well on continuous optimization problems. In this paper, we investigate its use in discrete optimization and particularly on the well known Travelling Salesman Problem (TSP). This investigation concerns the implementation of the idea of short and long runners when searching for Hamiltonian cycles in complete graphs. The approach uses the notion of k-optimality. The performance of the algorithm on a standard list of test problems is compared to that of the Genetic Algorithm (GA), Simulated Annealing (SA), Particle Swarm Optimization (PSO) and the New Discrete Firefly Algorithm (New DFA). Computational results are included.

Keywords Discrete optimization · Plant Propagation Algorithm · Heuristics · Nature-inspired algorithms · Travelling Salesman Problem

1 Introduction

Real world problems are often hard to solve. Nature has evolved ways to deal with problems over millions of years. Today, when standard mathematical methods do not work, Scientists turn to Nature for inspiration. Hence the rapid development in the so called Nature-inspired algorithms or heuristics, [8]. Heuristics, generally, do not guarantee to find the optimum solution. However, they are often able to find good approximate solutions, in reasonable time, [55]. Some of the well-known algorithms in this class are the Genetic Algorithm [15], Simulated Annealing [23], Tabu Search [10], Ant Colony Optimization [6] and Particle Swarm Optimization [22]. Despite a long list of such algorithms, the complexity of problems that arise in daily applications, their size and the need to solve them in near real-time means that more robust and more efficient algorithms are required, [41]. Examples of algorithms which have been introduced recently are the Artificial Bee Colony algorithm, [19],

B.İ. Selamoğlu (✉) · A. Salhi
University of Essex, Wivenhoe Park, Colchester CO4 3SQ, UK
e-mail: bisela@essex.ac.uk

A. Salhi
e-mail: as@essex.ac.uk

© Springer International Publishing Switzerland 2016
X.-S. Yang (ed.), *Nature-Inspired Computation in Engineering*,
Studies in Computational Intelligence 637, DOI 10.1007/978-3-319-30235-5_3

43

the Firefly Algorithm [54], Cuckoo Search [57], the Bat Algorithm [56] and the Game Theory-based Multi-Agent approach, [44].

This paper is concerned with the Plant Propagation Algorithm or PPA, which implements the way the strawberry plant propagates; PPA is also referred to as the Strawberry Algorithm, [41]. It has been shown to work well on continuous unconstrained and constrained optimization problems, [47–49]. Here, we intend to establish that it can also work on discrete optimization problems. The specific problem under consideration is the Travelling Salesman Problem (TSP).

The rest of the paper is organised as follows. In Sect. 2, the TSP is defined and briefly reviewed. PPA is explained in Sect. 3 and its extension to handle discrete optimization problems is given in Sect. 4. Computational results and conclusion are given in Sect. 5.

2 TSP: A Brief Review

Combinatorial optimization is frequently met in real applications. Combinatorial or discrete optimisation problems are often computationally demanding. They more often than not belong to the so called NP-hard class of problems, [3, 9]. As such, it is not reasonable to expect exact solutions when solving large and practical instances. Thus, approximation methods, heuristics and meta-heuristics, are almost the norm when it comes to solving them, [35, 53].

A notoriously difficult and yet easy to state representative of this class of problems is the well known Travelling Salesman Problem, or TSP, [25]. The aim is to find the Hamiltonian cycle of shortest length in the complete weighted graph that represents a fully connected set of cities where the edges represent the connections between each pair of cities; an edge weight is the distance (time, cost ...) that separates a pair of cities linked by the edge. Depending on the properties of these weights, we get different types of TSP. When the weights $c_{i,j}$ are put together in a square matrix $C = c_{i,j}, \forall\, i, j$ and $c_{i,j} = c_{j,i}, \forall\, i, j$, then we have a symmetric TSP or STSP, by virtue of matrix C which is symmetric. It is asymmetric if this property does not hold. If entries of C fulfil the triangle inequality, i.e. $c_{i,k} \leq c_{i,j} + c_{j,k}, \forall\, i, j, k$, the TSP is called metric. When $c_{i,j}$ is given as the Euclidean distances between nodes, the TSP is said to be Euclidean, [4, 25].

There are various methods to solve the TSP. As for other intractable combinatorial optimisation problems, exact algorithms are available, but only for relatively small instances of TSP; they do not work for large instances for efficiency reasons. Therefore, many heuristic approaches have been proposed in this respect, [14, 40].

Some of the exact algorithms for TSP are the Held-Karp algorithm which is based on dynamic programming [12], branch and bound [24], and branch and cut algorithms [33]. Some of the popular heuristic algorithms are Lin-Kernighan Local Search [27] and the strip algorithm [51]. Metaheuristics have become popular since they are able to find near optimal results in reasonable time even for large instances [38, 43, 58]. These are Nature-inspired algorithms which are now widely used to solve both continuous and discrete optimization problems. Most of these have been applied to TSP. Early examples are the Genetic Algorithm (GA) [11], Simulated Annealing (SA) [1], Ant Colony Optimisation (ACO) [7], Discrete Particle Swarm Optimisation (DPSO) [5, 45] and Tabu Search (TS) [30].

In the following we review some of the most recently introduced Nature-inspired algorithms. Tsai et al. [52] have developed an algorithm called Heterogeneous Selection Evolutionary Algorithm (HeSEA). The algorithm has been tested on 16 TSP benchmark problems ranging from 318 to 13509-city problems. The algorithm is able to find the optimum result for up to 3038-city problem. The average errors are 0.05, 0.01 and 0.74 % for problems that have 4461, 5915 and 13509 cities, respectively. Song and Yang [46] have proposed an improved ACO using dynamic pheromone updating and a mutation strategy in order to increase the quality of the original algorithm. The new approach has given better results than the classical ACO and has found even better results than the best known solutions for some TSPs. Marinakis et al. [31] have proposed a hybrid algorithm to solve the Euclidean TSP problem. Their approach combines Honey Bees Mating Optimization algorithm (HBMOTSP), the Multiple Phase Neighborhood Search-Greedy Randomized Adaptive Search Procedure (MPNS-GRASP) and the Expanding Neighborhood Search Strategy. Experiments have been run on 74 benchmark TSP instances and have given competitive results.

Karaboga and Gorkemli [20] implemented the Combinatorial Artificial Bee Colony algorithm (CABC). They have adapted the Greedy Sub Tour Mutation (GSTM) operator proposed by Albayrak and Allahverdi [2], which increases the capability of GA to find the shortest length in TSP. The algorithm was used to solve two TSP instances with 150 and 200 cities, respectively. In [21], Gorkemli et al. have introduced the Quick Combinatorial Artificial Bee Colony Algorithm (qCABC) and improved CABC by changing the behaviour of onlooker bees. The new algorithm was tested against 9 heuristic methods including the CABC on same instances. The qCABC outperforms all algorithms except CABC on the 150-city problem. In [26], Li et al. have developed a Discrete Artificial Bee Algorithm (DABC) and applied it to TSP. They used a Swap Operator to represent the basic ABC for discrete problems. The performance of the algorithm was compared to that of PSO algorithm. Experimental results show that DABC outperforms PSO.

Jati and Suyanto [17] have introduced the Evolutionary Discrete Firefly Algorithm (EDFA) and tested it against the Memetic Algorithm (MA). EDFA was found to be better than MA on TSP instances. An improved version of EDFA was developed later by Jati et al. [16]; it uses a new movement scheme. This new version has outperformed the previous one in terms of efficiency.

Ouaarab et al. have introduced the Improved Cuckoo Search (ICS) and also two types of Discrete Cuckoo Search (DCS) algorithms, one of which is based on the original CS and the other one on the improved CS, to solve the TSP. The performance of the improved DCS was tested on 41 problems with the number of cities ranging from

Table 1 A compilation of recent notable results

Authors	Year	Algorithm	TSP size	Avg. error (%)
Song et al.	2006	Improved ACO	48–250 cities	0 % for all problems
Marinakis et al.	2011	HBMOTSP	51–85,900 cities	0 % for 63 out of 74 instances, less than 1 % for 11 instances
Karaboga et al.	2011	CABC	150 and 200 cities	0.9 % and 0.6 % respectively
Li et al.	2011	DABC	14–130 cities	Changing from 0.55 to 6.41 %
Jati et al.	2011	EDFA	16–666 cities	0 % up to 225-city instances and the 666-city problem, less than 12 % for instances of 225, 280 and 442 cities
Karaboga et al.	2013	qCABC	150 and 200 cities	0.7 % and 0.5 % respectively
Jati et al.	2013	New EDFA	16–666 cities	0 % up to 225-city instances and the 666-city problem, less than 12 % for instances of 225, 280 and 442 cities
Ouaarab et al.	2014	DCS	51–1379 cities	0 % for 13 out of 41 instances, less than 4.78 % as the worst for the 1379-city instance
Saenphon et al.	2014	FOGS-ACO	48–200 cities	0 % for the instance of 51-city, changing from 0.062 to 1.64 % for the other instances
Mahi et al.	2015	PSO-ACO-3Opt	51–200 cities	Changing from 0.00 to 0.95 %

51 to 1379. The experimental results show that the proposed method is superior to Genetic Simulated Annealing Ant Colony System with Particle Swarm Optimization Technique (GSA-ACS-PSOT) and DPSO [32]. Saenphon et al. [39] have developed the Fast Opposite Gradient Search method and combined it with ACO. The proposed method has been compared to TS, GA, PSO, ACO, PS-ACO and GA-PS-ACO on TSP instances. Mahi et al. have developed an algorithm based on PSO, ACO and 3-opt algorithms for TSP, [28]. The new hybrid method was tested against some well-known algorithms. The experimental results show that it outperforms the other

algorithms in terms of solution quality. The results of the new algorithm were considered to be satisfactory. Table 1 is a compilation of a set of results reported in fairly recent papers. The last column records the performance of the concerned algorithm on a set of TSP problems.

3 The Basic Plant Propagation Algorithm

The Plant Propagation Algorithm (PPA) introduced by Salhi and Fraga, [41], emulates the strategy that plants deploy to survive by colonising new places which have good conditions for growth. Plants, like animals, survive by overcoming adverse conditions using strategies. The strawberry plant, for instance, has a survival and expansion strategy which is to send short runners to exploit the local area if the latter has good conditions, and to send long runners to explore new and more remote areas, i.e. to run away from a not so favourable current area.

Algorithm 1 Pseudo-code of PPA, [41]

1: Generate a population $P = X_i$, $i = 1, \ldots, NP$;
2: $g \leftarrow 1$
3: **for** $g = 1 : g_{\max}$ **do**
4:     Compute $N_i = f(X_i), \forall\, X_i \in P$
5:     Sort P in ascending order of fitness values N (for minimization);
6:     Create new population Φ
7:     **for** each X_i, $i = 1, \ldots, NP$ **do**
8:         $r_i \leftarrow$ set of runners where both the size of the set and the distance for each runner (individually) are proportional to the fitness values N_i;
9:         $\Phi \leftarrow \Phi \cup r_i$ (append to population; death occurs by omission);
10:    **end for**
11:    $P \leftarrow \Phi$ (new population);
12: **end for**
13: **return** P, (the population of solutions).

The algorithm starts with a population of plants each of which represents a solution in the search space. X_i denotes the solution represented by plant i in an n-dimensional space. $X_i \in R^n$, i.e. $X_i = [x_{i,j}]$, for $j = 1, \ldots, n$ and $x_{ij} \in R$. NP is the population size, i.e. $i = 1, \ldots, n$ where $n_{\max}$ denotes the maximum number of runners that each plant can send. This iterative process stops when g the counter of generations reaches its given maximum value $g_{\max}$.

Individuals/plants/solutions are evaluated and then ranked (sorted in ascending or descending order) according to their objective (fitness) values and whether the problem is a min or a max problem. The number of runners of a plant is proportional to its objective value and conversely, the length of each runner is inversely proportional to the objective value, [41]. For each X_i, $N_i \in (0, 1)$ denotes the normalized objective function value. The number of runners for each plant to generate is

$$n_r^i = \lceil (n_{\max} \, N_i \, \beta_i) \rceil \tag{1}$$

where n_r^i shows the number of runners and $\beta_i \in (0, 1)$ is a randomly picked number. For each plant, the minimum number of runners is set to 1. The distance value found for each runner is denoted by dx_j^i. It is:

$$dx_j^i = 2(1 - N_i)(r - 0.5), \; for \; j = 1, \dots, n. \tag{2}$$

where $r \in [0, 1]$ is a randomly chosen value.

Calculated distance values are used to position the new plants as follows:

$$y_{i,j} = x_{i,j} + (b_j - a_j) \, dx_j^i, \; for \; j = 1, \dots, n. \tag{3}$$

where $y_{i,j}$ shows the position of the new plant and $[a_j, b_j]$ are the bounds of the search space.

The new population that is created by appending the new solutions to the current population is sorted. In order to keep the number of population constant, the solutions that have lower objective value are dropped.

4   Extension to Discrete Optimization Problems

PPA has been shown to work well on continuous unconstrained and constrained optimization problems, [47–49]. In this study, we consider the case of the Travelling Salesman Problem (TSP). The issues with the implementation of PPA to solve discrete optimization problems are:

1. Finding/Defining the equivalent of a distance between two solutions in the solution space which is a set of permutations representing tours.
2. Defining the neighbourhood of a solution, here a tour or permutation.

4.1   Implementation of PPA to Handle TSP

In any algorithm and in particular in population-based ones, representation of individuals/solutions is a key aspect of their implementation. The issue here is the representation of a plant which itself represents a solution. A solution here is any Hamiltonian cycle (tour) of the complete graph representation of the TSP. Note that representation affects the way the search/optimisation process as well as any stopping criteria are implemented.

4.1.1 The Representation of a Tour

A plant in the population of plants maintained by PPA is a tour/solution represented as a permutation of cities. X_i is tour i, $i = 1, \ldots, NP$. This means that the size of the population of plants is NP. Tours/plants are ranked according to their lengths. The tour length of plant i is denoted by N_i; it is a function of X_i. Without loss of generality, the Euclidean TSP is considered here. Tour lengths, therefore, are calculated according to the Euclidean distance

$$d_{x,y} = \sqrt{\sum_{i=1}^{n}(x_i - y_i)^2}. \tag{4}$$

In Fig.1, the entries of the array represent cities. City 1 and city 4 are successive in the depicted tour and the notation 1–4 defines the edge between them.

4.1.2 Distance Between Two Plants

One of the issues in implementing PPA is defining the distance that separates tours. Here it is defined as the number of exchanges to transform one tour into another. After sorting the tours by their tour lengths, a pre-determined number of the tours is taken amongst the ones that have good short lengths; short runners are then sent from these plants, i.e. new neighbouring tours are generated from them. The 2-opt rule is used for this purpose since it require the minimum number of changes to create new tours. The 2-opt move is implemented by removing two edges from the current tour and exchanging them with two other edges, [18].

An illustration of a 2-opt exchange can be seen in Fig.2. There, tour **a-b-d-c-a** has been transformed into **a-d-b-c-a** by exchanging edges **a-b** and **d-c**.

1	4	6	9	12	20	5	7	8	11	15	3	17	16	2	10	14	18	13	19

Fig. 1 The permutation representation of a plant as a tour

Fig. 2 2-opt exchange of a 4-city tour

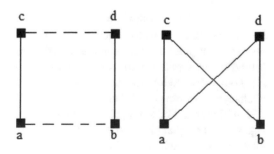

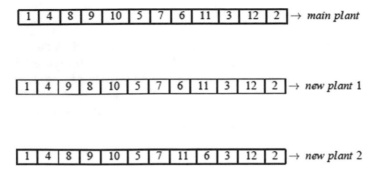

Fig. 3 Illustration of new plants generated by sending short runners from the main plant

Similarly, long runners are implemented by applying a k-opt rule with $k > 2$. In fact, this is pretty much the Lin-Kernighan algorithm (LK). It changes k edges in a tour, with k other edges. If, in this process, shorter tours are preferred and kept, then it will converge to potentially better solutions than it started with, [13].

This is not the only way available to measure the distance separating any two tours or permutations. However, it is in our view the best given the its efficiency and ease of implementation. An alternative approach can be found in [29], where a metric space of permutations defined using the k-opt rule has been studied.

4.1.3   Short and Long Runners

As mentioned earlier, in the basic PPA, a plant sends many short runners when it represents a good solution (exploitation move), or a few long runners when representing a poor solution (exploration move). Short runners are implemenetd using the 2-opt rule, and long runners, the k-opt rule, with $k > 2$.

An illustration of 2-opt rule implementing short runners can be seen in Fig. 3. In the figure, only two 2-opt neighbours of the main plant are shown. The first new plant was generated by exchanging the edges **4-8** and **8-9** and for the second one the edges **7-6** and **6-11** were exchanged. Note that the exchanged edges do not have to be adjacent.

Those tours in the population deemed to be representing poor solutions send one long runner each to explore the search space for better solutions. This is reasonable since a plant in a poor spot can hardly afford to send many long runners. Here, the long distance separating tours is implemented by applying a k-opt rule with $k > 2$. For both short and long runner cases, if the new tours have better results from these exchanges they are adopted and kept as new tours. Otherwise they are ignored.

An illustration of a new plant produced by sending a long runner can be seen in Fig. 4. The new plant is a 6-opt neighbour of the main plant. A 6-opt move can either be achieved by exchanging 6 edges chosen with other 6 edges or by implementing a number of 2-opt moves sequentially.

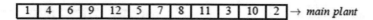

Fig. 4 Illustration of new plants generated by sending 1 long runner from the main plant

4.2   Pseudo-code of Discrete PPA

To the light of the general idea of implementing discrete PPA, we aimed at keeping the total computation time as short as possible. Therefore, it has been decided to start the algorithm with a good population of plants (tours). Diversity in the initial population is assumed to be guaranteed by the random processes used to generate tours. There are various such processes. Here, the initial population is generated using the greedy algorithm, random permutation or the strip algorithm, [38].

Algorithm 2 Pseudo-code of Discrete PPA

1: Generate a population $P = X_i$, $i = 1, \ldots, NP$ of valid tours; choose values for g_{max} and y.
2: $g = 1$
3: **while** g $< g_{max}$ **do**
4:     Compute $N_i = f(X_i), \forall\, X_i \in P$
5:     Sort $N = N_i$, $i = 1, \ldots, NP$ in ascending order (for minimization);
6:     **for** $i = 1 : E(NP/10)$, Top 10 % of plants **do**
7:         Generate $\lceil (y/i) \rceil$ short runners for plant i using 2-opt rule, where y is an arbitrary parameter.
8:         **if** $N_i > f(r_i)$ **then**
9:             $X_i \leftarrow r_i$
10:       **else**
11:           Ignore r_i
12:       **end if**
13:     **end for**
14:     **for** $i = E(NP/10) + 1 : NP$ **do**
15:         $r_i = 1$ runner for plant i using k-opt rule, $k > 2$, 1 long runner for each plant not in the top 10 percent
16:         **if** $N_i > f(r_i)$ **then**
17:             $X_i \leftarrow r_i$
18:         **else**
19:             Ignore r_i
20:         **end if**
21:     **end for**
22: **end while**
23: **return** P, (the population of solutions).

Strip heuristics are preferred when solving large TSP instances, because they are simple to implement and relatively fast, [38]. The strip approach can be explained

Fig. 5 An illustration of the
Classical Strip Algorithm

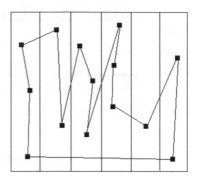

as dividing the plane into vertical or horizontal strips, then grouping cities in each strip according to the order of their coordinates, ascending in some and descending in others, alternating between the two in the process. The length of the tour is calculated using the sum of the Euclidean distances between cities including that between the first and the last, [50]. An illustration of the strip algorithm can be seen in Fig. 5.

For short runners, the 2-opt algorithm has been implemented using speed-up techniques such as, "don't look bits", and "fixed radius search". Details of these mechanisms can be found in [18, 36]. For long runners, the 2-opt has been implemented sequentially many times in order to complete the number of exchanges a k-opt rule with $k > 2$ would achieve, [13].

5 Computational Results and Discussion

Discrete PPA has been implemented using the 2-opt rule to represent the concept of a short runner and a sequence of 2-opt moves to implement that of a long runner. Three sets of experiments have been carried out. The first set compares PPA with GA and SA. The second compares it with Modified PSO and the third with the New DFA. Each of these is discussed below.

5.1 PPA Versus GA and SA

The discrete implementation of PPA has been applied to 10 TSP instances ranging from 14 to 101 cities, [37]. These well known problem have also been solved with GA and SA elsewhere, in particular, in [42, 44]. The key parameters used in our experiments can be found in Table 2, below. For algorithms GA and SA, the relevant parameter values can be found in Tables 4 and 5.

Two termination criteria can be used: the first is the maximum number of generations set to 100, and the second is the number of iterations without any change in the

Table 2 Parameters used in PPA experimental results

Problem size	Population size	Max. gen[a]	Short runners	Long runners
14–51	40	100	10 (each a 2-opt)	1 (3 seq. 2-opts)
51–102	40	100	10 (each a 2-opt)	1 (4 seq. 2-opts)
102–666	100	100	10 (each a 2-opt)	1 (6 seq. 2-opts)

[a] Maximum number of generations
seq. = sequential

Table 3 Comparison between GA, SA and PPA for solving standard TSP instances, [37]

Problem	Optimum	GA		SA		Discrete PPA	
		Av. Dv. (%)	Av. Time (s)	Av. Dv. (%)	Av. Time (s)	Av. Dv. (%)	Av. Time (s)
burma14	30.8785	0.7	6.95	0.53	8.34	0	1.55
ulysses16	73.9876	0.27	8.26	0.17	42.28	0	2.71
ulysses22	75.3097	1.56	9.83	1.16	97.12	0	4.02
att48	33524	4.97	41.23	31.48	10.52	0.6	5.71
eil51	426	4.46	44.45	18.17	423.26	1.54	5.12
berlin52	7542	8.67	42.99	36.37	11.44	2.1	7.87
st70	675	11.62	66.17	24.89	232.21	1.66	9.23
eil76	538	6.84	74.9	33.34	1162.02	4.1	9.42
pr76	108,159	6.25	94.02	35.91	254.2	1.2	10.26
eil101	629	10.37	143.99	50.11	220.71	4.29	14.37

Table 4 Parameters of genetic algorithm

Population size	50
Maximum number of generations	20
The rate of crossover	0.95
The rate of mutation	0.075
The length of the chromosome	50 × 8-bits
The number of points of crossover	2

best tour found so far, which is set to 10. In these experiments, PPA outperformed both GA and SA on all instances. All algorithms have been coded in Matlab and each algorithm was run 5 times. Results of the comparison with GA and SA have been recorded in Table 3. Note that although the results are very encouraging, further testing on larger instances and comparison with other algorithms are needed to draw useful conclusions on performance. The really interesting aspects of PPA which have not been discussed yet are: (a) it is very simple to understand; (b) it involves less parameters than GA and SA, for instance.

Claim (a), above, is justified if we note that the algorithm is based on a universal principle which is "to stay in a favourable spot" (exploitation) and "to run away from an unfavourable spot" (exploration). That is all that the algorithm implements really.

Table 5 Parameters of
simulated annealing

Maximum temperature	20
Minimum temperature	1
$\alpha\%$	0.5%
P	5
Number of iterations at each temperature	20
Temperature set	[20,10,5,3,1]

But, that is all a global search algorithm requires too. Claim (b) is equally easy to justify. Let us consider the list of parameters that are arbitrarily set in GA.

1. The population size;
2. The maximum number of generations;
3. The number of generations without improvement to stop;
4. The rate of crossover;
5. The rate of mutation;
6. The length of the chromosome;
7. The number of points of crossover.

Now, compare the above list to that of PPA. We can start the algorithm with a single plant that will then produce more plants unlike GA where a population with more than one individual is required. The number of runners can be decided by the objective value of each plant; indeed in Nature, some plants may have no runners at all because they are in desperate conditions while other may have 1, 2 or more depending on where they are and the corresponding value they give to the objective function. If we accept this, then PPA requires no more than a mechanism to stop, i.e. a stopping criterion. Hence, the comparatively short list of its parameters as in Table 4.

1. The maximum number of generations;
2. The number of generations without improvement to stop;
3. The maximum number of runners any plant can have.

For ease of implementation more parameters are used.

Simulated Annealing is not as extravagant as GA when it comes to arbitrary parameters. However, it still requires at least five parameters.

1. The maximum temperature;
2. The minimum temperature;
3. Parameter α: Percentage improvement in the objective value expected in each move;
4. The maximum number of moves without achieving $\alpha\%$ of objective function value improvement (P);
5. Number of iterations at each temperature.

To this, one can add the temperature set as in the last row of Table 5.

It is, therefore, fair to say that PPA compares well against GA and SA even if only small instances of TSP have been considered. Note that the proliferation of arbitrary parameters makes the concerned algorithms less usable since it is difficult to find good default parameters when the list is long. More arbitrary parameters also mean more uncertainty. Based on this comparison approach, it is fair too to say that PPA will match most heuristics and hyper-heurlstics. Further work may be to design a more realistic algorithm comparison methodology which not only takes into account raw performance, but also what is required in terms of parameter setting.

5.2   PPA Versus Modified PSO

A second set of experiments has been conducted and the results compared to those obtained by the Modified PSO, [34]. There are four versions of PSO was studied. All algorithms have been applied to 4 TSP instances with 14 to 76 cities, [37]. Each algorithm was run 10 times for each problem. The results of the comparison can be found in Table 6. The parameters values used in the PPA experiments can be found in Table 2. Those of PSO can be found in Table 7.

Table 6 Comparison between modified PSO and PPA for solving standard TSP instances, [37]

Problem	PSO-TS	PSO-TS-2opt	PSO-TS-CO	PSO-TS-CO-2opt	Discrete PPA
	Av. Dv. (%)	Av. Dv. (%)	Av. Dv. (%)	Av. Dv. (%)	Av. Dv. (%)
burma14	9.12	0	10	0	0
eil51	35.47	6.81	16.34	2.54	1.84
eil76	9.98	5.46	12.86	4.75	3.76
berlin52	7.37	5.22	10.33	2.12	1.84

PSO-TS : The PSO based on space transformation
PSO-TS-2opt: PSO-TS combined with 2-opt local search
PSO-TS-CO: PSO-TS with chaotic operations
PSO-TS-CO-2opt: PSO-TS combined with CO and 2-opt

Table 7 Parameters of modified PSO

The number of particles	50
The value of V_{max}	0.1
The values for learning factors, c1 and c2	$c1 = c1 = 2$
The inertia coefficient	1
P_{max}	1
Local search probability	0.01
Dispitive probability	0.001
The maximum number of generations	2000

In these experiments, PPA outperformed all Modified PSO algorithms on all instances in terms of solution quality. Arbitrarily set parameters for Modified PSO are as listed below, [34]:

1. The number of particles;
2. The value of V_{max};
3. The values for learning factors, c1 and c2;
4. The inertia coefficient;
5. A positive real number P_{max}, to express the range of the activities for each particle;
6. Local search probability;
7. Dispitive Probability;
8. The maximum number of generations;
9. The number of generations without improvement to stop.

Figure 6a–d depict the tours found in generations 1, 3, 5 and 8 of PPA when applied to a 22-city instance [37]. The last figure shows the optimal tour. Figure 7a–d show the convergence curves of the algorithm on four TSP instances with the number of cities ranging from 48 to 225.

5.3 PPA Versus New DFA

Another set of experiments has been conducted and the results compared to those obtained by the Discrete Firefly Algorithm (New DFA) described in [16, 58]. Both algorithms have been applied to 7 TSP instances with 16 to 666 cities, [37]. The parameter values used in PPA experiments can be found in Table 2. For New DFA, they can be found in Table 9. Note that the average accuracy was calculated using the equation below.

$$Avg.\ accuracy = \frac{Best\ known\ solution}{Avg.\ solution\ found} * 100 \qquad (5)$$

Each algorithm was run 50 times. The results of the comparison can be found in Tables 8 and 9.

In these experiments, in terms of solution quality PPA outperformed New DFA on 3 out of 7 instances. On the remaining 4 instances both algorithms have found the optimum solution. Arbitrarily set parameters required by New DFA are as listed below, [16]:

1. The number of maximum generations;
2. The population size;
3. The light absorption coefficient;
4. The updating index.

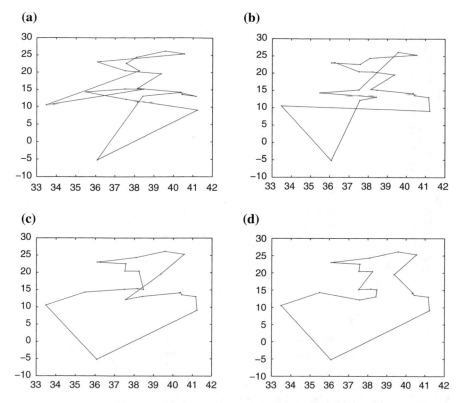

Fig. 6 **a** 22-city problem-1st Generation. **b** 22-city problem-3rd Generation. **c** 22-city problem-5th Generation. **d** 22-city problem-8th Generation

5.4 Conclusion and Further Investigation

We have shown that PPA can be implemented to handle discrete problems and in particular the TSP. The results, although limited given the sizes of the problems considered, are very encouraging and overall in favour of PPA. The idea of comparing algorithms/heuristics on other criteria than just performance is appealing in many respects; the criterion consisting of the number of arbitrarily set parameters required by a given algorithm/heuristic is interesting because it does not depend on the programming skills of the researcher, the quality of codes, the language of coding, the compiler, the processor etc. It is intrinsic to the algorithm/heuritic. Further research, therefore, could be to explore how the number of arbitrarily set parameters required by different algorithms/heuristics, affects their performance and distinguishes between them. It is also interesting to know if a parameter is at all necessary in the implementation of a heuristic. In the case of PPA, the number of plants is not really necessary to start with since the algorithm works perfectly well starting with a single plant.

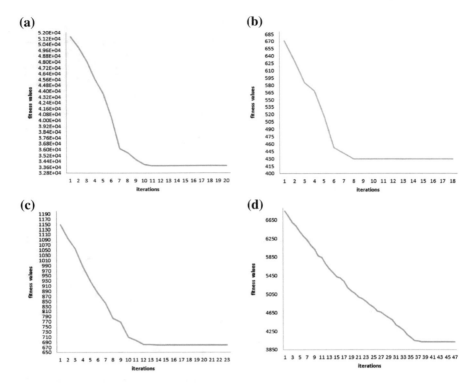

Fig. 7 **a** Curve evolution diagram of att48. **b** Curve evolution diagram of eil51. **c** Curve evolution diagram of st70. **d** Curve evolution diagram of tsp225

Table 8 PPA versus new DFA and PPA on standard TSP instances, [37]

Problem	Optimum	New DFA	PPA
		Av. Acc. (%)	Av. Acc. (%)
ulysses16	73.9876	100	100
ulysses22	75.3097	100	100
gr202	549.99	100	100
tsp225	3845	88.332	94.242
a280	2578	88.297	93.392
pcb442	50778	88.505	93.985
gr666	3952.53	100	100

Table 9 Parameters of new DFA

The number of maximum generations	Between 100–500 generations
The population size	5
The light absorption coefficient	0.001
The updating index	Between 1 and 16

Similar insights can be gained if this is considered when experimenting with a variety of heuristics.

Further investigation concerns extending PPA to solve other discrete problems such as, Knapsack, set covering and facility location problems. Results of this extension will be reported in future papers.

References

1. Aarts, E.H., Korst, J.H., van Laarhoven, P.J.: A quantitative analysis of the simulated annealing algorithm: a case study for the traveling salesman problem. J. Stat. Phys. **50**(1–2), 187–206 (1988)
2. Albayrak, M., Allahverdi, N.: Development a new mutation operator to solve the traveling salesman problem by aid of genetic algorithms. Expert Syst. Appl. **38**(3), 1313–1320 (2011)
3. Arora, S., Lund, C., Motwani, R., Sudan, M., Szegedy, M.: Proof verification and the hardness of approximation problems. J. ACM **45**(3), 501–555 (1998). doi:10.1145/278298.278306. http://doi.acm.org/10.1145/278298.278306
4. Bellmore, M., Nemhauser, G.L.: The traveling salesman problem: a survey. Oper. Res. **16**(3), 538–558 (1968)
5. Clerc, M.: Discrete particle swarm optimization, illustrated by the traveling salesman problem. In: New Optimization Techniques in Engineering, pp. 219–239. Springer (2004)
6. Dorigo, M.: Ant colony optimization and swarm intelligence. In: Proceedings of 5th International Workshop, ANTS 2006, Brussels, Belgium, September 4–7, 2006, vol. 4150. Springer Science & Business Media (2006)
7. Dorigo, M., Gambardella, L.M.: Ant colonies for the travelling salesman problem. BioSystems **43**(2), 73–81 (1997)
8. Fister Jr., I., Yang, X.S., Fister, I., Brest, J., Fister, D.: A brief review of nature-inspired algorithms for optimization. arXiv preprint arXiv:1307.4186 (2013)
9. Garey, M.R., Johnson, D.S.: Computers and Intractability: A Guide to the Theory of NP-Completeness. W. H. Freeman & Co., New York (1979)
10. Glover, F., Laguna, M.: Tabu Search. Springer (1999)
11. Grefenstette, J., Gopal, R., Rosmaita, B., Van Gucht, D.: Genetic algorithms for the traveling salesman problem. In: Proceedings of the First International Conference on Genetic Algorithms and their Applications, pp. 160–168. Lawrence Erlbaum, New Jersey (1985)
12. Held, M., Karp, R.M.: A dynamic programming approach to sequencing problems. J. Soc. Ind. Appl. Math 196–210 (1962)
13. Helsgaun, K.: An effective implementation of k-opt moves for the lin-kernighan tsp heuristic. Ph.D. thesis, Roskilde University. Department of Computer Science (2006)
14. Hoffman, K.L., Padberg, M., Rinaldi, G.: Traveling salesman problem. In: Encyclopedia of Operations Research and Management Science, pp. 1573–1578. Springer (2013)
15. Holland, J.H.: Adaptation in Natural and Artificial Systems: An Introductory Analysis with Applications to Biology, Control, and Artificial Intelligence. U Michigan Press (1975)
16. Jati, G.K., Manurung, R., Suyanto: 13–discrete firefly algorithm for traveling salesman problem: a new movement scheme. In: X.S.Y.C.X.H.G. Karamanoglu (ed.) Swarm Intelligence and Bio-Inspired Computation, pp. 295–312. Elsevier, Oxford (2013). doi:10.1016/B978-0-12-405163-8.00013-2. http://www.sciencedirect.com/science/article/pii/B9780124051638000132
17. Jati, G.K., Suyanto: Evolutionary Discrete Firefly Algorithm for Travelling Salesman Problem. In: Bouchachia, Abdelhamid (ed.) Adaptive and Intelligent SystemsSpringer, pp. 393–403.Springer, Berlin, Heidelberg (2011). ISBN:978-3-642-23857-4

18. Johnson, D.S., McGeoch, L.A.: The traveling salesman problem: a case study in local optimization. Local Search Comb. Optim. **1**, 215–310 (1997)
19. Karaboga, D., Basturk, B.: A powerful and efficient algorithm for numerical function optimization: artificial bee colony (abc) algorithm. J. Global Optim. **39**(3), 459–471 (2007)
20. Karaboga, D., Gorkemli, B.: A combinatorial artificial bee colony algorithm for traveling salesman problem. In: Innovations in Intelligent Systems and Applications (INISTA), 2011 International Symposium on, pp. 50–53. IEEE (2011)
21. Karaboga, D., Gorkemli, B.: A quick artificial bee colony-qabc-algorithm for optimization problems. In: Innovations in Intelligent Systems and Applications (INISTA), 2012 International Symposium on, pp. 1–5. IEEE (2012)
22. Kennedy, J.: Particle swarm optimization. In: Encyclopedia of Machine Learning, pp. 760–766. Springer (2010)
23. Kirkpatrick, S., Gelatt, C.D., Vecchi, M.P.: Optimization by simulated annealing. Science **220**(4598), 671–680 (1983)
24. Land, A.H., Doig, A.G.: An automatic method of solving discrete programming problems. Econometrica: J. Econometric Soc. 497–520 (1960)
25. Laporte, G.: The traveling salesman problem: an overview of exact and approximate algorithms. Eur. J. Oper. Res. **59**(2), 231–247 (1992)
26. Li, L., Cheng, Y., Tan, L., Niu, B.: A discrete artificial bee colony algorithm for tsp problem. In: Bio-Inspired Computing and Applications, pp. 566–573. Springer (2012)
27. Lin, S., Kernighan, B.W.: An effective heuristic algorithm for the traveling-salesman problem. Oper. Res. **21**(2), 498–516 (1973). doi:10.1287/opre.21.2.498. http://dx.doi.org/10.1287/opre.21.2.498
28. Mahi, M., mer Kaan Baykan, Kodaz, H.: A new hybrid method based on particle swarm optimization, ant colony optimization and 3-opt algorithms for traveling salesman problem. Appl. Soft Comput. 30, 484–490 (2015). doi:10.1016/j.asoc.2015.01.068. http://www.sciencedirect.com/science/article/pii/S1568494615000940
29. Mak, K.T., Morton, A.J.: Distances between traveling salesman tours. Discrete Appl. Math. **58**(3), 281–291 (1995). doi:10.1016/0166-218X(93)E0115-F. http://www.sciencedirect.com/science/article/pii/0166218X93E0115F
30. Malek, M., Guruswamy, M., Pandya, M., Owens, H.: Serial and parallel simulated annealing and tabu search algorithms for the traveling salesman problem. Ann. Oper. Res. **21**(1), 59–84 (1989)
31. Marinakis, Y., Marinaki, M., Dounias, G.: Honey bees mating optimization algorithm for the euclidean traveling salesman problem. Inform. Sci. **181**(20), 4684–4698 (2011)
32. Ouaarab, A., Ahiod, B., Yang, X.S.: Improved and discrete cuckoo search for solving the travelling salesman problem. In: Cuckoo Search and Firefly Algorithm, pp. 63–84. Springer (2014)
33. Padberg, M., Rinaldi, G.: A branch-and-cut algorithm for the resolution of large-scale symmetric traveling salesman problems. SIAM Rev. **33**(1), 60–100 (1991)
34. Pang, W., Wang, K.P., Zhou, C.G., Dong, L.J., Liu, M., Zhang, H.Y., Wang, J.Y.: Modified particle swarm optimization based on space transformation for solving traveling salesman problem. In: Proceedings of 2004 International Conference on Machine Learning and Cybernetics, 2004, vol. 4, pp. 2342–2346. IEEE (2004)
35. Papadimitriou, C.H., Steiglitz, K.: Combinatorial Optimization: Algorithms and Complexity. Courier Corporation (1998)
36. Rego, C., Gamboa, D., Glover, F., Osterman, C.: Traveling salesman problem heuristics: leading methods, implementations and latest advances. Eur. J. Oper. Res. **211**(3), 427–441 (2011)
37. Reinelt, G.: TSPLIB–A T.S.P. library. Tech. Rep. 250, Universität Augsburg, Institut für Mathematik, Augsburg (1990)
38. Reinelt, G.: The Traveling Salesman: Computational Solutions for TSP Applications. Springer, Berlin (1994)
39. Saenphon, T., Phimoltares, S., Lursinsap, C.: Combining new fast opposite gradient search with ant colony optimization for solving travelling salesman problem. Eng. Appl. Artif. Intell. **35**,

324–334 (2014). doi:10.1016/j.engappai.2014.06.026. http://dx.doi.org/10.1016/j.engappai. 2014.06.026

40. Salhi, A.: The ultimate solution approach to intractable problems. In: Proceedings of the 6th IMT-GT Conference on Mathematics, Statistics and its Applications, pp. 84–93 (2010)

41. Salhi, A., Fraga, E.: Nature-inspired optimisation approaches and the new plant propagation algorithm. In: Proceedings of the ICeMATH2011 pp. K2–1 to K2–8 (2011)

42. Salhi, A., Proll, L.: Rios Insua, D., Martin, J.: Experiences with stochastic algorithms for a class of constrained global optimisation problems. RAIRO–Oper. Res. 34, 183–197 (2000). doi:10.1051/ro:2000110

43. Salhi, A., Rodríguez, J.A.V.: Tailoring hyper-heuristics to specific instances of a scheduling problem using affinity and competence functions. Memetic Comput. 6(2), 77–84 (2014)

44. Salhi, A., Töreyen, Ö.: A game theory-based multi-agent system for expensive optimisation problems. In: Computational Intelligence in Optimization, pp. 211–232. Springer (2010)

45. Shi, X.H., Liang, Y.C., Lee, H.P., Lu, C., Wang, Q.: Particle swarm optimization-based algorithms for tsp and generalized tsp. Inform. Process. Lett. 103(5), 169–176 (2007)

46. Song, X., Li, B., Yang, H.: Improved ant colony algorithm and its applications in tsp. In: Sixth International Conference on Intelligent Systems Design and Applications, 2006. ISDA'06, vol. 2, pp. 1145–1148. IEEE (2006)

47. Sulaiman, M., Salhi, A.: A Seed-based plant propagation algorithm: the feeding station model. Sci World J (2015)

48. Sulaiman, M., Salhi, A., Fraga, E.S.: The Plant Propagation Algorithm: Modifications and Implementation. ArXiv e-prints (2014)

49. Sulaiman, M., Salhi, A., Selamoglu, B.I., Kirikchi, O.B.: A plant propagation algorithm for constrained engineering optimisation problems. Mathematical Problems in Engineering 627416, 10 pp (2014). doi:10.1155/2014/627416

50. Supowit, K.J., Plaisted, D.A., Reingold, E.M.: Heuristics for weighted perfect matching. In: Proceedings of the Twelfth Annual ACM Symposium on Theory of Computing, pp. 398–419. ACM (1980)

51. Supowit, K.J., Reingold, E.M., Plaisted, D.A.: The travelling salesman problem and minimum matching in the unit square. SIAM J. Comput. 12(1), 144–156 (1983)

52. Tsai, H.K., Yang, J.M., Tsai, Y.F., Kao, C.Y.: An evolutionary algorithm for large traveling salesman problems. IEEE Trans. Syst. Man Cybern. Part B: Cybern. 34(4), 1718–1729 (2004)

53. Yagiura, M., Ibaraki, T.: On metaheuristic algorithms for combinatorial optimization problems. Syst. Comput. Jpn 32(3), 33–55 (2001)

54. Yang, X.S.: Firefly algorithm, stochastic test functions and design optimisation. Int. J. Bio-Inspir. Comput. 2(2), 78–84 (2010)

55. Yang, X.S.: Nature-inspired Metaheuristic Algorithms, 2nd edn. (2010)

56. Yang, X.S.: A new metaheuristic bat-inspired algorithm. In: Nature Inspired Cooperative Strategies for Optimization (NICSO 2010), pp. 65–74. Springer (2010)

57. Yang, X.S., Deb, S.: Cuckoo search via lévy flights. In: World Congress on Nature & Biologically Inspired Computing, 2009. NaBIC 2009. pp. 210–214. IEEE (2009)

58. Yang, X.S., Cui, Z., Xiao, R., Gandomi, A.H., Karamanoglu, M.: Swarm Intelligence and Bio-inspired Computation: Theory and Applications. Newnes (2013)

Enhancing Cooperative Coevolution with Surrogate-Assisted Local Search

Giuseppe A. Trunfio

Abstract In recent years, an increasing effort has been devoted to the study of metaheuristics suitable for large-scale global optimization in the continuous domain. However, so far the optimization of high-dimensional functions that are also computationally expensive has attracted little research. To address such an issue, this chapter describes an approach in which fitness surrogates are exploited to enhance local search (LS) within the low-dimensional subcomponents of a cooperative coevolutionary (CC) optimizer. The chapter also includes a detailed discussion of the related literature and presents a preliminary experimentation based on typical benchmark functions. According to the results, the surrogate-assisted LS within subcomponents can significantly enhance the optimization ability of a CC algorithm.

Keywords Large scale global optimization · Evolutionary optimization · Differential evolution · Memetic algorithms · Cooperative coevolution · Surrogate fitness

1   Introduction

In recent years, nature-inspired optimization metaheuristics have been successfully applied to a number of real-world problems. However, advancements in science and technology increasingly challenge existing algorithms with new complex applications. Among the latter, there are optimization problems characterized by search spaces with several hundreds to thousands of variables.

Unfortunately, optimization metaheuristics suffer from the so-called *curse of dimensionality* [4], that is, even if they show excellent search capabilities when applied to small or medium sized problems, their performance decays rapidly as the dimensionality of the problem increases. Therefore, the study of more effective and efficient search strategies to address high-dimensional optimization problems, especially with constraints on the available computational budget, has been recently

G.A. Trunfio (✉)
DADU, University of Sassari, Alghero, Italy
e-mail: trunfio@uniss.it

© Springer International Publishing Switzerland 2016   63
X.-S. Yang (ed.), *Nature-Inspired Computation in Engineering*,
Studies in Computational Intelligence 637, DOI 10.1007/978-3-319-30235-5_4

recognized as a relevant field of research, which is often referred to as 'large-scale global optimization' (LSGO) [42, 50].

In the field of continuous optimization, various approaches have been proposed to effectively address LSGO problems by improving the scalability of search algorithms. Among the most successful techniques, it is worth mentioning the hybridization of different search operators [39, 41, 88], the use of adaptive algorithms [8, 39, 41, 59, 92, 95], the use of local search (LS) [37, 39, 41, 53, 54] and the cooperative coevolutionary (CC) approach [65, 66]. The latter is based on the divide-and-conquer principle, which performs a decomposition of the search space in order to obtain sub-problems of smaller size. Although many of the developed strategies were quite effective in addressing LSGO problems, there is still room for significant contributions in the field [42, 50].

For example, so far little research has been attracted by the issue of optimizing high-dimensional functions that are also computationally expensive. When fitness evaluation is difficult, a typical approach in optimization is to exploit cheaper approximations of the objective function during the search (i.e., the so-called *surrogates* or *meta-models*) [28, 30]. However, for high-dimensional problems also the quality of fitness surrogates is plagued by the curse of dimensionality. One approach for addressing such a problem consists of using surrogates in the context of the CC framework. Although this technique has been already proposed in the literature [23, 61], so far its use in LSGO has not been thoroughly studied.

Among the several ways to exploit surrogates within optimization algorithms, there is the enhancement of LS [28, 30]. Given that the latter characterizes some of the most effective algorithms for LSGO [37, 39, 41, 53, 54], it looks promising to investigate a CC optimization algorithm in which a LS phase is assisted by fitness surrogates. Some preliminary results of such a study are discussed in this chapter, which is organized as follows.

Section 2 presents a quite detailed discussion of the related literature, including the CC approach, relevant algorithms that exploit LS to improve their scalability, surrogate-assisted optimization and its application to CC. Then, Sect. 3 describes the proposed approach and Sect. 4 discusses some preliminary results. The chapter ends with Sect. 5, in which some directions of future research are outlined.

2 Related Work

According to the literature, several techniques have been used so far for effectively addressing LSGO problems [42, 50]. An important family of LSGO methods is based on the CC approach [65, 66], in which the original high-dimensional problem is decomposed into lower-dimensional sub-components.

Another typical enhancement of optimization algorithms is based on *Local Search* (LS) [37, 39, 41, 53, 54], which consists of attempting to improve some selected individuals of the population by searching their neighbourhoods through suitable heuristics.

Among the most successful approaches, it is also worth mentioning the *hybridization* of different techniques, which was often a key factor for developing highly scalable optimization algorithms [39, 41, 88]. Moreover, different forms of *adaptation* were successfully adopted for improving the behaviour of search algorithms [8, 57, 58, 67, 93, 95, 96]. Avoiding premature convergence is another aspect that has been explicitly addressed by some successful LSGO algorithms. Typically, this is obtained through a diversity control mechanism triggered by suitable indicators (e.g., the reduction of variance in the population) [13]. Recently, also the issue of addressing very high-dimensional decision spaces (e.g., millions of variables) was tackled using high performance computing jointly applied with some techniques mentioned above [37].

In the following, we provide some details on LSGO algorithms and other techniques that are relevant to the approach proposed in this chapter.

2.1 Cooperative Coevolution

CC is a divide-and-conquer technique in which the original high-dimensional problem is decomposed into lower-dimensional easier-to-solve *subcomponents* [65]. Typically, each subcomponent is solved using an ordinary population-based optimization metaheuristics. During the process, the only cooperation takes place in the fitness evaluation through an exchange of information between subcomponents.

CC was first applied to a GA by Potter and De Jong in [65]. Subsequently, the approach has been successfully tested with many search algorithms such as Ant Colony Optimization [17], Particle Swarm Optimization [18, 43, 63, 87], Simulated Annealing [74], Differential Evolution (DE) [92], Firefly Algorithm [83] and many others.

In practice, a CC optimization is based on partitioning the d-dimensional set of search directions $G = \{1, 2, \ldots, d\}$ into k sets $G_1 \ldots G_k$. Each group G_i of directions defines a subcomponent whose dimension can be significantly lower than d. By construction, a candidate solution found by a subcomponent contains only some elements of the d-dimensional vector required for computing the corresponding fitness function f. Thus, to evaluate the latter, a common d-dimensional *context vector* **b** is built using a representative individual (e.g., the best individual) provided by each subcomponent. Then, before its evaluation, each candidate solution is complemented through the appropriate elements of the context vector. In this framework, the cooperation between sub-populations emerges because the common vector is used for the fitness evaluation of all individuals.

In the typical decomposition of the original d-dimensional search space into k subcomponents of the same dimension $d_k = d/k$, the groups of directions associated to the subcomponents are defined as:

$$G_i = \{(i-1) \times d_k + 1, \ldots, i \times d_k\} \tag{1}$$

and the context vector is:

$$\mathbf{b} = (\underbrace{b_1^{(1)}, \ldots, b_{d_k}^{(1)}}_{\mathbf{b}^{(1)}}, \underbrace{b_1^{(2)}, \ldots, b_{d_k}^{(2)}}_{\mathbf{b}^{(2)}}, \ldots, \underbrace{b_1^{(k)}, \ldots, b_{d_k}^{(k)}}_{\mathbf{b}^{(k)}})^T \qquad (2)$$

where $\mathbf{b}^{(i)}$ is the d_k-dimensional vector representing the contribution of the i-th subcomponent (e.g., its current best position):

$$\mathbf{b}^{(i)} = (b_1^{(i)}, b_2^{(i)}, \ldots, b_{d_k}^{(i)})^T \qquad (3)$$

Given the j-th individual $\mathbf{x}^{(i,j)} \in S^{(i)}$ of the i-th subcomponent:

$$\mathbf{x}^{(i,j)} = (x_1^{(i,j)}, x_2^{(i,j)}, \ldots, x_{d_k}^{(i,j)})^T \qquad (4)$$

its fitness value is given by $f(\mathbf{b}^{(i,j)})$, where $\mathbf{b}^{(i,j)}$ is defined as:

$$\mathbf{b}^{(i,j)} = (\underbrace{b_1^{(1)}, \ldots, b_{d_k}^{(1)}}_{\mathbf{b}^{(1)}}, \ldots, \underbrace{x_1^{(i,j)}, \ldots, x_{d_k}^{(i,j)}}_{\mathbf{x}^{(i,j)}}, \ldots, \underbrace{b_1^{(k)}, \ldots, b_{d_k}^{(k)}}_{\mathbf{b}^{(k)}})^T \qquad (5)$$

In other words, the fitness of $\mathbf{x}^{(i,j)}$ is evaluated on the vector obtained from $\mathbf{b}$ by substituting the components provided by the i-th sub-population with the corresponding components of $\mathbf{x}^{(i,j)}$.

Except for the evaluation of individuals, the optimization is carried out using a standard optimizer in each subcomponent.

2.1.1 Random Grouping

A major issue with the CC approach is that, when interdependent variables are assigned to different subcomponents, the search efficiency can decline significantly [65, 66]. In fact, with the *static* decomposition method outlined above, interdependent variables are likely to be located in different subcomponents for the whole optimization process. For this reason, some effective decomposition methods have been developed for dealing with the case of partially separable problems [12, 25, 57, 58, 76], that is, problems in which subsets of interacting variables can be recognized and grouped together. Nevertheless, many large-scale optimization problems are fully non-separable. For such cases, a typical decomposition approach consists of using equally sized subcomponents together with the Random Grouping (RG) strategy proposed by Yang et al. in [92, 93]. In RG the directions of the original search space are periodically grouped in a random way to determine the CC subcomponents. Such an approach was successfully integrated into several CC optimizers [43, 63, 83].

Algorithm 1: CCRG(objectiveFunction, d)

1   $\mathcal{G} = \{G_1, \ldots, G_k\} \leftarrow$ randomGrouping(d, k);
2   $pop \leftarrow$ initPopulation($d, numInd$);
3   $\mathbf{b} \leftarrow$ initContextVector(pop);
4   $FE \leftarrow 0$;
5   **while** *fitnessEvaluations* $<$ *maxFE* **do**
6 | $\langle f, FE \rangle \leftarrow$ optimizeSubcomponents(objectiveFunction, $\mathcal{G}, \mathbf{b}, pop$);
7 | *fitnessEvaluations* $\leftarrow$ *fitnessEvaluations* $+ FE$;
8 | $\mathcal{G} = \{G_1, \ldots, G_k\} \leftarrow$ randomGrouping(n, k);
9 | //*depending on the optimizer further operations may be required after RG*
10   return f and $\mathbf{b}$;

Algorithm 2: optimizeSubcomponents(objectiveFunction, $\mathcal{G}, \mathbf{b}, pop$)

1   **foreach** $G_i \in \mathcal{G}$ **do**
2 | $pop_i \leftarrow$ extractPopulation(pop, G_i);
3 | $\langle pop_i, best_i, FE \rangle \leftarrow$ optimizer(objectiveFunction, $pop_i, \mathbf{b}, G_i, numIte$);
4 | $pop \leftarrow$ storePopulation(pop_i, G_i);
5 | $\mathbf{b} \leftarrow$ updateContextVector($best_i, G_i$);
6   return objectiveFunction($\mathbf{b}$), FE;

More in details, in the linear decomposition proposed in [46, 65, 87] the i-th sub-population operates on the group of directions G_i defined by Eq. (1). In addition, the decomposition is static, in the sense that it is defined before the beginning of optimization cycles. Instead, a RG approach assigns to the i-th group $d_k = d/k$ directions q_j, with j randomly selected without replacement from the set $\{1, 2, \ldots, d\}$.

The typical CC implementation endowed with the RG strategy is shown in Algorithms 1 and 2. The first step in Algorithm 1 consists of creating k groups of coordinates randomly drawn without replacement from the set $\{1, 2, \ldots, d\}$, where d is the dimension of the problem. Then, both the population, composed of *numInd* individuals, and the context vector are randomly initialized. The optimization is organized in *cycles*. During each cycle, as shown in Algorithm 2, the optimizers are activated in a round-robin fashion for the different subcomponents and the context vector is updated using the current best individual of each sub-population. A budget of *numIte* optimizer iterations is allocated to each subcomponent at each cycle. At the end of the latter, a new random decomposition is carried out. The CC cycles terminate when the number of fitness evaluations reaches the value *maxFE*.

2.1.2   Successful Algorithms Based on Cooperative Coevolution

Among the most successful CC implementations, there is the *DECC-G*, originally presented by Yang et al. in [92] and based on the 'Self-Adaptive with Neighborhood

Search Differential Evolution' (SaNSDE) algorithm [94]. The latter is a self-adaptive DE in which the mutation operator is replaced by a random neighbourhood search. The DECC-G algorithm was the runner-up at the 2013 CEC special session on LSGO. Effective evolutions of DECC-G are the *MLCC* and *MLSoft* algorithms [59, 93], where the size of subcomponents is adaptively selected through a rein-forcement learning approach.

Another relevant CC algorithm is the *CC-CMA-ES* [45], which applies CC to CMA-ES in order to scale up CMA-ES [24] to large-scale optimization. This algo-rithm obtained the third position at the 2013 CEC special session on LSGO.

It is also worth mentioning the 'Two-stage based ensemble optimization for Large-Scale Global Optimization' (*2S-Ensemble*) algorithm, which was proposed in [88, 89]. It divides the search procedure into two different stages: (i) in the first stage, a search technique with high convergence speed is used to shrink the search region on a promising area; (ii) in the second stage, a CC based search technique is used to exploit such a promising area to get the best possible solution. The CC uses ran-domly three different optimizers, based on SaDE (an adaptive DE algorithm) [67], on a GA and a standard DE. Moreover, the size of the decomposition is adaptive. The 2S-Ensemble algorithm was the runner-up at the 2010 CEC special session on LSGO.

2.2 Memetic Algorithms for LSGO

As mentioned above, an effective way to improve the efficiency of optimization metaheuristics consists of hybridizing the process with LS techniques. The latter are typically applied to some selected and promising individuals at each generation. In such a search process, commonly referred to as Memetic Algorithm (MA) [55, 56], a global search mechanism helps to explore new search zones, while the LS exploits some promising solutions.

Clearly, since LS can be computationally expensive, a good balance between exploration (main algorithm) and exploitation (LS) is a key factor for the success of a MA implementation. This is particularly true in case of problems with high dimensionality. In fact, in such cases LS concerns much challenging neighbourhoods and a suitable value of its *intensity* (i.e. the number of fitness function evaluations assigned to LS) should be carefully determined in order to achieve a satisfactory search efficiency.

A successful evolutionary MA, which was very effective on the 20 test problems proposed for the LSGO special session in the CEC 2010 [78], is the *MA with LS chains* (MA-LS-Chains), proposed in [52, 53] and later developed in [54]. In brief, at each generation the LS chain method resumes, on some individuals, the LS exactly from its last state (i.e., that resulting from the LS in the previous generation). Such an approach allows to effectively adapt LS parameters, including its intensity, during the search process.

The most effective MA-LS-Chains, named MA-SW-Chains, exploits as LS the Solis and Wets' algorithm [75], which is an adaptive randomized hill-climbing heuristic. The SW process starts with a solution $\mathbf{x}^{(0)}$ and explores its neighbourhood, through a step-by-step process, to find a sequence of improved solutions $\mathbf{x}^{(1)}$, $\mathbf{x}^{(2)} \ldots \mathbf{x}^{(q)}$. More in details, in SW two neighbours are generated at each step by adding and subtracting a random deviate $\boldsymbol{\Delta}$, which is sampled from a normal distribution with mean $\mathbf{m}$ and standard deviation ρ. If either $\mathbf{x}^{(i)} + \boldsymbol{\Delta}$ or $\mathbf{x}^{(i)} - \boldsymbol{\Delta}$ is better than $\mathbf{x}^{(i)}$, the latter is updated and a *success* is registered. Otherwise, the value of $\boldsymbol{\Delta}$ is considered as a *failure*. On the basis of the number of successes, and depending on some fixed parameters, the values of ρ and $\mathbf{m}$ are updated during the process, in order to both increase the convergence speed and bias the search towards better areas of the neighbourhood. The process continues up to a certain number of fitness function evaluations (i.e., the LS intensity).

An important characteristic of MA-SW-Chain is that, to avoid super exploitation, the total number of fitness function evaluations dedicated to LS is a fixed fraction of the total available budget. In addition, at each generation only one individual is chosen to be improved by LS using a strategy that allows activating new promising chains and to exploit existing ones.

Other hybrid algorithms endowed with LS that were very successful in dealing with LSGO problems are those based on the Multiple Offspring Sampling (MOS) approach [38], in which different mechanisms for creating new individuals are used in the optimization. During the process, the goodness of each involved mechanism is evaluated according to some suitable metrics. Then, the latter are used to dynamically adapt the participation, in terms of computational effort, of each technique.

A MOS algorithm that showed the best performance on the test set proposed for the 2011 special issue of *Soft Computing* journal on LSGO [39], was based on Differential Evolution (DE) and a LS strategy. The algorithm is composed of a fixed number of steps. At each step, a specific amount FEs of fitness function evaluations is distributed between the involved techniques $T^{(i)}$ according to some *Participation Ratios* Π_i, which are computed accounting for the quality of the individuals produced at the previous time step.

As for the adopted LS, referred to as MTS-LS1, it was based on one of the methods included in the Multiple Trajectory Search (MTS) algorithm [85]. In brief, it searches separately along each direction using a deterministic search range (*SR*) initialized to a suitable value SR_0. The value of *SR* is reduced to one-half if the previous LS does not lead to improvements. When *SR* falls below a small threshold, it is reinitialized to SR_0. Along each search direction, the solution's coordinate is first subtracted by *SR* to look for fitness improvements. In case of improvement, the search proceeds to consider the next dimension. Otherwise, the variable corresponding to the current direction is restored and then is added by $SR/2$, again to check if the fitness improves. If it is, the search proceeds to consider the next dimension. If it is not, the variable is restored and the search proceeds to consider the next direction. The detailed pseudo-code can be found in [85].

Another MOS-based hybrid algorithm was later presented at the CEC 2012 LSGO session, where it outperformed all the competitors [40]. In this case, the approach combined two LS techniques without a population-based algorithm, namely the MTS-LS1 [85] and the SW algorithm [75] already outlined above. Such an approach, besides achieving the best result at the CEC 2012 special session on LSGO, in a comparison combining the results of CEC 2010 and CEC 2012 sessions, also out-performed many other algorithms [79].

Later, a new MOS-based hybrid algorithm, which combines a GA with two strategies of LS [41], was the best performing algorithm in the test set proposed for LSGO session at the CEC 2013 (i.e., a set of 15 large-scale benchmark problems, with dimension up to 1000, devised as an extension to the previous CEC 2010 benchmark suite).

2.3 Surrogate-Assisted Optimization and Its Application to CC

In case of computationally expensive fitness functions, the efficiency of algorithms in exploiting the information available from past fitness evaluations can be significantly increased building a *surrogate model* (or a *meta-model*) $\hat{f}(\mathbf{x})$ of $f(\mathbf{x})$ [29].

Several techniques have been proposed in the literature to reduce the number of expensive fitness evaluations using cheap surrogates and to effectively exploit them within an optimization process. Typical suitable meta-models include Polynomial Regression (PR) [10, 16, 22, 100], Gaussian Processes (GPs) [19, 44, 47, 64, 86, 90, 97], Artificial Neural Networks (ANNs) [2, 15, 26, 27, 72, 90] and Radial Basis Function Networks (RBFNs) [16, 70, 80, 91].

The basic idea of meta-modelling is that building and using the surrogate $\hat{f}$ is a relatively cheap process if compared with evaluating the exact fitness f. The model $\hat{f}$ is built using an appropriate learning algorithm, which exploits an archive $\mathscr{A}$ of couples $\langle \mathbf{x}, f(\mathbf{x}) \rangle$ obtained during or before the optimization process.

Typically, the learning algorithm optimizes an objective function with respect to a suitable metric of the distance between $\hat{f}$ and f. However, as pointed out by Jin in [30], defining the optimality of $\hat{f}$ is not straightforward. In fact, the quality of surrogate models for optimization algorithms is not necessarily a close quantitative approximation of the original fitness function (e.g., obtained by minimizing the mean squared error). For example, when the surrogate is used to support the selection of individuals, its ability to produce a correct ranking of offspring can be much more important than approximation accuracy. Unfortunately, as shown empirically by Jin et al. in [33] the mean square error of the model only weakly correlates with the ability to select correct individuals. In addition, sometimes a smoother surrogate is preferred to a more accurate approximation of a rugged fitness function f, because of its ability to prevent the search from getting trapped in local optima (such an effect was referred to as 'blessing of uncertainties' in [62]).

That said, being able to estimate the approximation error can be helpful in developing adaptive and more efficient surrogate-assisted algorithms. A first approach to estimation of model quality is *re-sampling* during the learning process, that is, using part of the available dataset to test the model, as outlined later in this section. A simpler approach presented in the literature consists of assuming that the model quality decreases with the distance from the points included in the training dataset [7]. However, some meta-modelling technique also provides an estimate of the uncertainty, in terms of variance. This is the case of GPs [49], which have increasingly been adopted as surrogate in optimization algorithms [9, 14, 44, 64, 97], in spite of the fact that the computational cost of learning can be high in case of a large number of samples. As detailed later, also the application presented in this chapter is based on GPs.

2.3.1 Training Meta-models

Broadly speaking, the surrogate model $\hat{f}$ can be either used on the whole decision space or it can be specifically trained for being applied on a limited sub-domain. In the first case of *global surrogate*, $\hat{f}$ is built less frequently during the search, in particular if a specific adaptive mechanism of surrogate quality control is implemented [30]. The second case involves a greater computational cost because many local surrogates must be built for different regions of the search space (e.g., a local surrogate can be trained ad-hoc for evaluating a new individual, or groups of individuals close to each other). Often, especially in case of high-dimensional and multimodal fitness landscapes, an accurate global approximation of the true fitness function is difficult to obtain [30].

Two main approaches exist for building the surrogate fitness function. In particular, in the so-called *off-line learning* the model is built before the optimization starts, by exploiting existing data (e.g., patterns available from previous optimization runs). A different strategy, named *online learning*, consists of constructing the surrogate fitness model using data that are generated during the optimization process. The latter approach has been reported to be more successful and reliable than the former [90].

Moreover, also depending on the type of surrogate (i.e., local or global), on the meta-modelling technique (e.g., PR, RBFN, GP, etc.) and adopted learning algorithm, there are several strategies for maintaining and exploiting the archive $\mathscr{A}$ of past fitness evaluations. Optimal selection strategies for training and test data depend on both the used surrogate modelling technique and optimization algorithm.

A common approach consists of learning the surrogate on the k-nearest neighbour points to a given point $\bar{x}$. The latter can be a new individual to be evaluated (i.e., in the case of local surrogate) or some other average point representing a cluster of individuals. For example, using Euclidean distance as similarity measure, several clusters $\mathscr{A}_i$ can be computed from the elements of the archive $\mathscr{A}$. Then, an independent model $\hat{f}_i$ can be trained using the elements of each cluster $\mathscr{A}_i$. To evaluate a new individual x: first, the cluster $\mathscr{A}_k$ whose centroid is the closest to x is chosen; then, the value $\hat{f}_k(x)$ is associated to x [47].

Another strategy consists of using the k most recently evaluated points, which are expected to be closer to the current search area and, for this reason, should be able to provide better surrogates.

Also combinations of the above methods can be devised, that is, building local models with the k-nearest neighbours, which are selected among the most recently evaluated individuals.

Another important aspect concerns the way in which the selected samples are used to train the surrogate. In the line of principle, standard strategies of model selection and validation (e.g., split sample validation, cross-validation, bootstrap, etc.) can be used to improve the generalization ability of the surrogate. For example, in split sample validation the selected dataset $\mathscr{A}_i$ is randomly partitioned into a *training set* $\hat{\mathscr{A}}_i$ and a *test set* $\bar{\mathscr{A}}_i$, with $|\bar{\mathscr{A}}_i| \ll |\hat{\mathscr{A}}_i|$. During the model learning, the set $\hat{\mathscr{A}}_i$ is used for training while the set $\bar{\mathscr{A}}_i$ is used to estimate the current model quality. In this way, it is possible to stop an iterative training process (e.g., the ANNs back propagation algorithm) before overfitting. A problem with this approach is that it reduces the amount of samples actually available for training. More complex procedures are cross-validation and bootstrap, which involve the re-sampling of multiple subsets of the original dataset to be used alternatively for model training and validation [5, 81]. In general, since each training set corresponds to a new model learning, such methods can be computationally expensive (although the cost should be compared to that of the true fitness function). Also for this reason, in spite of being very effective in model quality estimation, re-sampling approaches had only minor success in optimization [81].

2.3.2 Surrogate Model Exploitation

Clearly, the key aspect of surrogate-assisted optimization is the way in which the model $\hat{f}$ is actually used within the search algorithm.

According to the literature, a fitness surrogate can be exploited to reduce the randomness in population initialization, cross-over and mutation, obtaining the so-called *informed operators* [1, 3, 68]. The latter can promote a quicker convergence of the algorithm towards regions with improved fitness. Often, in this case, even surrogates of low quality can be helpful since they can provide better results than a random guess [30].

A fitness surrogate can also be used as a replacement of the exact fitness for evaluating new individuals [7, 20, 21, 29, 31, 34, 35]. In this case, the immediate objective is to significantly increase the computational efficiency of the optimization process. However, not always an optimum for the surrogate model is also an optimum for the true fitness function. Therefore, to avoid convergence to false optima, the surrogate-assisted optimization should also use the exact fitness function. On the other hand, the involvement of the latter should be minimized due to its high computational cost. A trade-off is provided by a suitable *evolution control* strategy [29].

In particular, two different approaches are commonly adopted. The first is the *individual-based control*, which consists of using the true fitness function to evaluate some individuals produced at each generation (i.e., the *controlled individuals*). The second is a *generation-based control*, in which all the individuals of some generations (i.e., the *controlled generations*) are evaluated through the exact fitness function.

In the case of individual-based control, different strategies can be used to choose the controlled individuals. For example, the *best strategy* consists of assigning their exact fitness to some individuals that are the best according to the meta-model [31, 32]. In particular, ensuring that the best individuals are always evaluated with the true fitness allows avoiding false optima in case of elitism. However, in the literature many other re-evaluation strategies have been investigated. For example, another approach consists of clustering the population and re-evaluating with the exact fitness only a representative individual for each cluster [21, 34, 35]. Also, when the surrogate has an associate quality measure (e.g., GPs), another possibility is to re-evaluate individuals with a large degree of uncertainty in approximation [7, 20]. As a positive side effect, this approach allows enriching the archive of solutions with points that are potentially the most suitable for improving the quality of next surrogates.

2.3.3 Surrogate-Assisted Local-Search

Another opportunity offered by the availability of a fitness surrogate is to improve the efficiency of LS [16, 51, 82, 98]. For example, given a surrogate in analytical form, it is possible to locally improve an individual using classical gradient-based methods. At the end of LS, after a re-evaluation trough the true fitness function, the locally optimized individual can replace the original one according to a Lamarckian evolutionary approach [60]. The advantage in this case is that even a surrogate of bad quality does not involve a risk of misleading the global search process [30]. Moreover, the minimum cost consists of just one evaluation of the exact fitness for each individual subjected to the LS process.

However, more sophisticated approaches can be used, such as the *trust-region strategy* [11], at the cost of a greater number of true fitness evaluations. The classical trust-region approach is an iterative procedure based on a quadratic model in which a constrained optimum is solved at each step. However, in the context of fitness surrogate the quadratic approximation has often been replaced by more powerful local models (either RBFs or GPs).

The idea behind the trust region approach is that even using an accurate surrogate $\hat{f}$, the LS procedure may converge towards a point that does not represent an actual improvement of the starting point $\mathbf{x}$. Hence, at the cost of some more evaluations of the exact fitness, the LS iteratively operates on a region in which the accuracy of $\hat{f}$ is verified using f. In particular, if the accuracy of $\hat{f}$ is above a threshold then the region is expanded; conversely, if $\hat{f}$ has poor accuracy then the region is contracted. In practice, according to the classical trust-region approach, the LS is structured in a sequence of sub-problems as follows:

$$\min \hat{f}(\mathbf{x}^{(j)} + \mathbf{d}), \quad j = 0, 1, 2, \ldots, \lambda \tag{6}$$

$$\text{subject to } \|\mathbf{d}\| \le r^{(j)} \tag{7}$$

where $\mathbf{x}^{(j)}$ is the starting point of the j-th iteration (i.e., $\mathbf{x}^{(0)}$ is the individual to optimize), $\mathbf{x}^{(j)} + \mathbf{d}$ represents a point within the current trust-region radius $r^{(j)}$. Each optimization sub-problem with bound constraints can be solved through a gradient-based method. At the first sub-problem of LS, the radius $r^{(0)}$ can be initialized as the average distances between a certain number of the nearest neighbours of $\mathbf{x}^{(0)}$ (e.g., those used for training a local surrogate). Subsequently, the value of $r^{(j)}$ is determined for each of the following sub-problems based on a parameter $\omega^{(j)}$, which is computed at the end of each subproblem as follows:

$$\omega^{(j)} = \frac{f(\mathbf{x}^{(j)}) - f(\mathbf{x}_{\text{opt}}^{(j)})}{\hat{f}(\mathbf{x}^{(j)}) - \hat{f}(\mathbf{x}_{\text{opt}}^{(j)})} \tag{8}$$

where $\mathbf{x}_{\text{opt}}^{(j)}$ is the current constrained optimum. Then, the trust region is contracted or expanded for high or low values of $\omega^{(j)}$ respectively, according some empirical rule (e.g., that described in [99]). In addition, before next step the trust-region is centred at the new optimum $\mathbf{x}_{\text{opt}}^{(j)}$.

The LS process terminates when the maximum number of subproblems λ is reached. The latter parameter represents the *individual learning intensity*, which is the amount of computational budget in terms of exact fitness evaluations devoted on improving a single solution.

2.3.4 Exploiting Surrogates in Cooperative Coevolution

As noted in [61], for high-dimensional decision spaces the use of surrogate-assisted optimization is limited by the difficulties of learning suitable surrogates. In fact, besides the main search algorithm, the curse of dimensionality also affects the process of training the meta-model.

However, since the CC approach decomposes the problem into lower-dimensional subcomponents, it seems the ideal framework for exploiting the surrogate-assisted methodologies described above in the case of LSGO. In spite of this, so far only few researchers addressed the problem of integrating meta-modelling into CC optimizers. Below, two relevant examples are outlined.

In [61], Ong and co-authors used a surrogate-assisted approach with a co-evolutionary real-coded genetic algorithm, implementing a mixed generation-based and individual-based control. Initially, the surrogate is used in each subcomponent. At each surrogate-based generation of each optimizer (see Algorithm 2), m clusters are created using the k-means algorithm. Then, the cluster centres are evaluated with the exact fitness function and used as centres of a RBF, which is adopted as a global surrogate for each subcomponent. The RBF is used for evaluating the population, before applying the standard evolutionary operators. However, the best individual is re-evaluated with the exact fitness function. When the co-evolutionary search on the surrogate stalls, the algorithm switches to the exact fitness for the next generations. According to the results, such a surrogate-assisted CC-GA proved promising, although only two test problems were used, with 20 and 40 variables respectively [61].

More recently, Goh and co-authors in [23] described a surrogate-assisted memetic CC algorithm for constrained optimization problems. In the proposed approach, the surrogate is used only to support the LS phase. The meta-modelling is based on a local polynomial regression and the used search operator is a solver based on sequential quadratic programming. At the end of LS, the optimized individual is re-evaluated with the exact fitness. Although the proposed algorithm has proved effective in some constraint-optimization benchmark test functions, it was not investigated in high-dimensional problems.

However, as illustrated in Sect. 2.2, recent research in LSGO showed that LS can be very helpful in some optimization problems. For this reason, exploiting surrogates for enhancement of LS deserves to be thoroughly investigated.

3 A CC Algorithm with Surrogate-Assisted LS

This section describes a memetic CC algorithm in which the LS phases are assisted by surrogates based on GPs.

The optimizer used in the CC search is JADE [96], an adaptive DE in which the parameter adaptation was implemented by evolving the mutation factors and crossover probabilities based on their historical record of success.

The LS phases are carried out independently within each subcomponent at each cycle, that is, on search spaces of dimension $d_k = d/k$, being k the number of subcomponents. This allows overcoming the training issues related to the high dimensionality, obtaining fitness surrogates of better quality. On the other hand, such an approach involves training a significant number of surrogates during the optimization process. Thus, the obtained computational advantage should be quantified accounting for the actual expensiveness of the fitness function evaluations.

Given that the number of subcomponents can be significant, the LS intensity must be carefully selected in the proposed approach (i.e., the total LS intensity is the sum of all the k LS intensities at the subcomponent level). For this reason, at each cycle of the CC algorithm, LS is applied only on a limited number of individuals and with a small budget of exact fitness evaluations. In particular, we apply a LS refinement to the set B_k composed of the q best individuals of the current population.

We developed two versions of the surrogate-assisted memetic CC. In the first, labelled as *CCGSLS*, a global surrogate is trained in each subcomponent before applying LS to the elements of B_k. In this case, the training set $\mathscr{A}_k$ is composed of the m most recently evaluated points. To this purpose, a fixed size FIFO *deque* is used for implementing the archive of individuals evaluated with the exact fitness.

Algorithm 3: optimizeSubcomponentsLS(objectiveFunction, $\mathcal{G}$, **b**, *pop*)

---
1 **foreach** $G_i \in \mathcal{G}$ **do**
2    $pop_i \leftarrow$ extractPopulation(*pop*, G_i);
3    $\langle pop_i, best_i, \mathscr{A}_i, FE \rangle \leftarrow$ optimizer(objectiveFunction, pop_i, **b**, G_i, *numIte*);
4    $\langle pop_i, best_i, \mathscr{A}_i, FE \rangle \leftarrow$ localSearch(objectiveFunction, pop_i, **b**, G_i, $\mathscr{A}_i$);
5    $pop \leftarrow$ storePopulation(pop_i, G_i);
6   **b** $\leftarrow$ updateContextVector($best_i$, G_i);

7 **return** objectiveFunction(**b**), FE;

In the second version, labelled as *CCLSLS*, a local surrogate is trained for each element $\mathbf{x}_i \in B_k$ before applying LS. In this case, the training set is composed of the p nearest neighbours ($p \leq m$) of $\mathbf{x}_i$, taken from the above set $\hat{\mathscr{A}}_k$. The closeness between individuals is based on their mutual Euclidean distance in the d_k-dimensional decision space. Note that in this case we need to train $k \times q$ surrogates at each cycle.

Moreover, to investigate the advantages provided by the meta-modelling approach, we developed a further version of the memetic-CC, labelled as *CCSWLS* and based on the Solis and Wets' algorithm [75] described in Sect. 2.2.

The memetic CC follows Algorithms 1, except for the subcomponent optimization phase, which is carried out according to Algorithm 3. In the latter, at each cycle the standard optimizer of each subcomponent produces an updated archive $\mathscr{A}_i$ of past evaluations with the exact fitness. Then, $\mathscr{A}_i$ is used within the *localSearch* function to train the surrogate(s) and perform LSs. It is worth noting that also during LS the archive is updated, as described below.

3.1 Training the Surrogate GP

We used GPs as fitness meta-modelling approach. In a GP, the value of a function $f(\mathbf{x})$ at any location $\mathbf{x}$ is considered as a *random variable* whose realizations follow a jointly Gaussian distribution. In other words, a GP can be thought of as an extension of the standard Gaussian distribution to a distribution over functions defined on a d-dimensional domain [69] (i.e., the subcomponent search space in our case). As any Gaussian distribution, also a GP is fully determined by its mean and variance, which are generalised into a *mean function*, $m(\mathbf{x})$, and *covariance function*, $k(\mathbf{x}_i, \mathbf{x}_j)$, both expressed in terms of the considered points:

$$m(\mathbf{x}) = E[f(\mathbf{x})] \tag{9}$$
$$k(\mathbf{x}_i, \mathbf{x}_j) = E[(f(\mathbf{x}_i) - m(\mathbf{x}_i))(f(\mathbf{x}_j) - m(\mathbf{x})_j)] \tag{10}$$

where $k(\mathbf{x}_i, \mathbf{x}_j)$ is a suitable function that may introduce hyper-parameters to the GP and $m(\mathbf{x})$ is typically a constant value (i.e., zero in the following). The idea of GP-based regression is to use the above prior belief about $f(\mathbf{x})$ together with some empirical observations (i.e., the training set) in order to describe the conditional distribution $p(f(\mathbf{x}) \mid \mathbf{x})$. Such a posterior distribution is computed according to the Bayes' rule.

In practice, a GP surrogate can be computed using a training set composed of n d-dimensional points $\mathbf{X} = \{\mathbf{x}_1, \ldots, \mathbf{x}_n\}$ with the corresponding values of the fitness function $\mathbf{y} = \{f(\mathbf{x}_1), \ldots, f(\mathbf{x}_n)\}$. The latter, according to the GP approach, is considered as a sample of a multivariate Gaussian distribution with joint probability density $p(\mathbf{y} \mid \mathbf{X})$. When a new test point $\mathbf{x}$ is added (i.e., the point in which the value of f should be estimated), also the resulting vector $\{\mathbf{y}, f(\mathbf{x})\}$ is considered as a sample of the Gaussian joint probability density $p(\{\mathbf{y}, f(\mathbf{x})\} \mid \{\mathbf{X}, \mathbf{x}\})$. Using the Bayes' rule, it can be easily shown that the expected value of $f(\mathbf{x})$ is given by [69]:

$$\hat{f}(\mathbf{x}) = \mathbf{k}(\mathbf{x}, \mathbf{X})^T \mathbf{C}(\mathbf{X}, \mathbf{X})^{-1} \mathbf{y} = \mathbf{k}(\mathbf{x}, \mathbf{X})^T \boldsymbol{\alpha} \qquad (11)$$

where: $\mathbf{k}(\mathbf{x}, \mathbf{X})$ is the $n \times 1$ vector of covariances between the test point $\mathbf{x}$ and the k training points (i.e., $k_i = k(\mathbf{x}, \mathbf{x}_i)$); $\mathbf{C}(\mathbf{X}, \mathbf{X})$ is the $n \times n$ matrix of covariances between the n training points (i.e., $C_{ij} = k(\mathbf{x}_i, \mathbf{x}_j)$; $\mathbf{y}$ is the vector of function values on the training points, as defined above. It is worth noting that since only $\boldsymbol{\alpha}$ is dependent on the training set, in case of global surrogate, a single inversion of the matrix $\mathbf{C}$ allows the estimation of multiple fitnesses. The same does not apply to the posterior variance of $f(\mathbf{x})$, which can be expressed as [69]:

$$Var[f(\mathbf{x})] = k(\mathbf{x}, \mathbf{x}) - \mathbf{k}(\mathbf{x}, \mathbf{X})^T \mathbf{C}(\mathbf{X}, \mathbf{X})^{-1} \mathbf{k}(\mathbf{x}, \mathbf{X}) \qquad (12)$$

The above Eqs. (11) and (12) were derived under the hypothesis of $m(\mathbf{x}) = 0$, which is the prior assumption, that is, before considering the data included in the training set. However, this is generally not a problem as in the region close to the data, the computed posterior distribution accounts for the actual values of the observations in vector $\mathbf{y}$.

In the most common applications, the covariance function $k(\mathbf{x}_i, \mathbf{x}_j)$ between two values $f(\mathbf{x}_i)$ and $f(\mathbf{x}_j)$ is only dependent on the distance between $\mathbf{x}_i$ and $\mathbf{x}_j$. In particular, it is usually reasonable to assume that the covariance between two function values increases as corresponding inputs get closer to each other. Among the many covariance functions developed in the literature, the preliminary application presented in this chapter was based on a squared exponential covariance function with isotropic length scale:

$$k(\mathbf{x}_i, \mathbf{x}_j) = \alpha^2 \exp(-\frac{1}{2}(\mathbf{x}_i - \mathbf{x}_j)^T \Lambda^{-1} (\mathbf{x}_i - \mathbf{x}_j)) \qquad (13)$$

where the $d \times d$ matrix $\Lambda = diag(l^2, \ldots, l^2)$ is defined by the characteristic length scale l, and α is an overall scale parameter. In practice, if the distance between input points is small compared to the length scale l, the exponential term is close to one; with increasing distance the covariance exponentially decays to zero. The advantage of Eq. (13) is that it only depends on a two-dimensional vector of hyperparameters $\theta = \{\alpha, l\}$.

Before using the GP model, the hyperparameters θ are optimized in such a way that the log-likelihood $\log(p(\mathbf{y} \mid \mathbf{X}, \theta))$ is maximal. Since the model derivatives with respect to the hyperparameters can be easily computed [69], in the line of principle, any gradient based optimization algorithm can be used to obtain a suitable value of θ (however, also this learning process may present some pitfalls due to local minima in the log-likelihood function).

In the present application, the GP-based surrogates where implemented using the C++ library *libgp* proposed in [6]. Besides providing several types of covariance functions, for the training phase *libgp* exploits *Rprop*, which is fast gradient-based optimization technique originally designed for neural network learning [71].

3.2 Local Search on Subcomponents

In both *CCLSLS* and *CCGSLS*, starting from each individual $\mathbf{x} \in B_k$, the local search is carried out using a trust-region approach, which interleaves exact fitness evaluations with the use of GP surrogates. Differently from the strategy described in Sect. 2.3.3, we replace Eq. (7) by a box constraint. In particular, the trust-region is defined as a cuboid with size $\rho^{(i)}$ centred at the current optimum point $\mathbf{x}^{(i)}$.

As mentioned above, in *CCLSLS*, for each initial guess $\mathbf{x} \in B_k$, a local GP is created using the p-nearest neighbours extracted from the dataset of the m most recent points that have already been evaluated using the exact fitness. Instead, in *CCGSLS* all the trust region processes are applied using a global GP built using the current archive.

It is worth noting that, according to results from the literature for MAs applied to LSGO [39–41, 52–54], LS intensity λ (i.e., the number of trust region steps) should be small to avoid super exploitation during the search process. This is especially true in case of CC, where the LS process is applied to $q \times k$ individuals at each cycle and even a small λ results in a significant number of additional evaluations of the exact fitness.

Nevertheless, the new points evaluated during the LS process are added to the training set $\hat{\mathscr{A}}_k$, so that they can contribute to the improvement of surrogates in subsequent cycles of the optimization.

To apply the LS according to the above procedure, the gradient of the adopted GP surrogate can be easily computed by differentiating the GP posterior mean expressed by Eq. 11 with respect to the input point:

$$\frac{\partial \hat{f}}{\partial \mathbf{x}} = -\Lambda^{-1} \tilde{\mathbf{X}}^T \left(\mathbf{k}(\mathbf{x}, \mathbf{X}) \odot \boldsymbol{\alpha} \right) \tag{14}$$

where $\tilde{\mathbf{X}} = \{\mathbf{x} - \mathbf{x}_1, \ldots, \mathbf{x} - \mathbf{x}_n\}^T$ is a $n \times d$ matrix and $\odot$ denotes an element-wise product between the two $n \times 1$ vectors $\mathbf{k}$ and $\boldsymbol{\alpha}$.

In the present application, the local search step is implemented using the quasi-Newton BFGS algorithm within the box-constrained optimizer included in the *dlib* C++ machine learning library [36].

In the *CCSWLS* the LS phase at the subcomponent level is based on the Solis and Wets' algorithm, without the need of building a surrogate. However, also in this case low search intensity, in terms of fitness evaluations, is adopted.

4 Some Preliminary Results

To carry out a preliminary evaluation of the surrogate-assisted memetic CC algorithm described above, we used the eight benchmark functions listed in Table 1. For all problems we adopted the dimension $d = 1000$ and the optimum value f^* is zero. Functions f_1 and f_2 are separable and characterized by many local minima. Function f_3 leads to a multi-modal separable optimization problem. The shifted uni-modal function f_4 was taken from the benchmark set proposed in [48]. Function f_5 is a common uni-modal separable optimization problem. Functions f_6–f_8 were taken from those proposed for the CEC'08 special session on LSGO [77]. In particular, f_6 is uni-modal and separable while f_7 and f_8 are multi-modal and non-separable.

The two JADE parameters referred to as c and p in [96] were set to 0.1 and 0.05 respectively. Also, we used one JADE iteration per cycle and subcomponent. The total number of exact fitness evaluations was set to 3.0E06. For all functions, we used 500 subcomponents of size $d_k = 2$, each with a population of 30 individuals. However,

Table 1 Used benchmark problems with dimension $d = 1000$

	Function name	Definition	Domain
f_1	Rastrigin's function	$f_1 = 10d + \sum_{i=1}^{d}(x_i^2 - 10cos(2\pi x_i))$	$[-5, 5]^d$
f_2	Ackley's function	$f_2 = 20exp(-0.2\sqrt{\frac{1}{d}\sum_{i=1}^{d}x_i^2})$ $- \exp(\frac{1}{d}\sum_{i=1}^{d}cos(2\pi x_i)) + 20 + e$	$[-32, 32]^d$
f_3	Styblinski-Tang function	$f_5 = \frac{1}{2}\sum_{i=1}^{d}(x_i^4 - 16x_i^2 + 5x_i) + 39.16599d$	$[-5, 5]^d$
f_4	Shifted Bohachevsky	$f_4 = \sum_{i=1}^{d}(z_i^2 + 2z_{i+1}^2 - 0.3\cos(3\pi z_i)$ $- 0.4\cos(4\pi z_{i+1}) + 0.7), z = x - o$	$[-15, 15]^d$
f_5	Sum of different powers	$f_5 = \sum_{i=1}^{d}|x_i|^i$	$[-1, 1]^d$
f_6	Shifted sphere	From CEC 2008 LSGO test suite	$[-100, 100]^d$
f_7	Shifted Rosenbrock function	From CEC 2008 LSGO test suite	$[-100, 100]^d$
f_8	Shifted Griewank function	From CEC 2008 LSGO test suite	$[-600, 600]^d$

For all functions the optimum value is $f^* = 0.0$ (note that the CEC'08 functions were redefined in order to have $f^* = 0.0$). Function f_4 is part of the benchmark set proposed in [48]

it is worth noting that for most functions and depending on the optimizer, a different number of subcomponents and individuals can lead to better result [59, 84]. Different experiments were carried out by varying the number q of the best individuals used to initialize LS at each cycle and within each subcomponent. In particular, we adopted $q \in \{1, 0.1p_i, 0.25p_i, 0.5p_i, 0.75p_i, p_i\}$, where $p_i = |pop_i| = 30$ is the size of the population in each subcomponent. For example, with $q = 1$ only the best individual was used for initializing a LS in each subcomponent, which corresponds to k LSs at each cycle (i.e., to 500 LSs in our case). The training set $\mathscr{A}$ was composed of the most recent 30 individuals and only 10 of them were used for building the local surrogates in *CCLSLS*. To better distribute the LS intensity along the optimization process, in both *CCLSLS* and *CCGSLS* each LS was applied for a maximum of $\lambda = 2$ exact fitness evaluations. Instead, in *CCSWLS* we executed each instance of the Solis and Wets's algorithm for a maximum of eight exact fitness evaluations. The initial step of LS was computed as 0.1 multiplied by the average Euclidean distance between the points involved in the GP training. For each test problem, we carried out 25 independent optimizations.

Table 2 shows some statistics on the achieved results for both the standard CC and the memetic algorithms with $q = 15$ (i.e., LS applied the best half of the population). In order to find the best performing algorithm we used a series of pair-wise Mann-Whitney-Wilcoxon (MWW) tests with Holm correction [73]. Statistical tests are based on 5 % significance level. In Table 2, the statistically best results for each function are marked in bold.

In addition, Figs. 1 and 2 show the average convergence plots obtained during the experiments.

According to Table 2, for function f_1 *CCLSLS* always achieved the exact optimum of $f^* = 0$. However, also *CCGSLS* provided an average result that can be considered equivalent in practice. Both surrogate-assisted algorithms largely outperformed the standard CC and the memetic *CCSWLS*. Also the speed of convergence of the surrogate-assisted algorithms, which can be seen in Fig. 1, can be considered satisfactory, as in half of the available budget of fitness evaluations the achieved accuracy was below 10^{-6}.

For function f_2, all algorithms under comparison, except *CCSWLS*, provided satisfactory approximations of the optimum. However, the best result of $1.308E - 12$ was provided by *CCGSLS*. Noticeably, the latter was much better than the standard CC in terms of speed of convergence in the first half of the optimization process.

In the optimization of function f_3 all memetic algorithms achieved statistically equivalent results and outperformed the standard CC. Moreover, it is worth noting from Fig. 1 how the surrogate-assisted algorithms can provide a much higher speed of convergence in the early stages of the optimization process.

For function f_4 the standard CC provided the best average result of $4.952E - 24$. The surrogate assisted memetic CCs also provided on average accurate approximations of the optimum, that is, $1.275E - 13$ and $6.907E - 14$, for *CCGSLS* and *CCLSLS* respectively. Instead, in *CCSWLS* the effort devoted to LS proved unfruitful. Interestingly, *CCGSLS* and *CCLSLS* provided a fast convergence achieving their best result in almost one third of the available computational budged. However, they

Table 2 Achieved results with the algorithms under comparison

	CC		CCSWLS	
	Avg. (std. dev.)	Best	Avg. (std. dev.)	Best
f_1	4.378E+00 (1.651E+00)	2.605E+00	3.612E+01 (1.803E+00)	3.489E+01
f_2	1.552E−12 (7.471E−14)	1.443E−12	3.083E−02 (1.290E−03)	2.942E−02
f_3	8.541E+00 (6.902E+00)	3.430E−02	**3.764E−02** (2.369E−04)	3.743E−02
f_4	**4.952E−24** (5.207E−25)	4.332E−24	3.646E−01 (2.162E−02)	3.456E−01
f_5	2.884E−06 (2.334E−06)	2.387E−07	4.562E−06 (6.594E−06)	9.208E−08
f_6	5.722E−23 (1.303E−23)	4.607E−23	2.954E−02 (2.237E−03)	2.605E−02
f_7	1.102E+02 (2.093E+02)	7.772E−01	7.290E+03 (3.224E+02)	6.796E+03
f_8	1.485E−03 (2.961E−03)	4.219E−15	3.961E−03 (8.911E−04)	2.580E−03
	CCGSLS		CCLSLS	
	Avg. (std. dev.)	Best	Avg. (std. dev.)	Best
f_1	2.487E−15 (4.974E−15)	0.000E+00	**0.000E+00** (0.000E+00)	0.000E+00
f_2	**1.308E−12** (1.212E−13)	1.121E−12	1.643E−11 (5.212E−12)	1.257E−11
f_3	**3.430E−02** (5.165E−11)	3.430E−02	**3.430E−02** (5.091E−11)	3.430E−02
f_4	1.275E−13 (3.108E−14)	9.161E−14	6.907E−14 (5.143E−14)	1.938E−15
f_5	2.080E−06 (2.919E−06)	1.036E−09	1.340E−06 (1.508E−06)	1.157E−07
f_6	**2.006E−36** (1.419E−36)	0.000E+00	3.070E−20 (4.341E−20)	0.000E+00
f_7	**8.622E+00** (2.839E+00)	2.943E+00	**9.104E+00** (1.935E+00)	5.233E+00
f_8	3.326E−02 (3.363E−02)	4.330E−15	**4.130E−15** (1.473E−16)	3.886E−15

At each cycle, the best 50 % of population (i.e., 15 individuals) was used for initializing a low-intensity LS

were not able to further improve the optimum approximation. The origin of such a phenomenon, and the way to prevent it, should be better investigated.

As can be seen from both Table 2 and Fig. 2, for function f_5 all the algorithms under comparison provided statistically equivalent results and behaved almost the same during the process. However, the achieved fitness, between 1.340E − 6 and 4.562E − 6, can be considered as acceptable in terms of accuracy. Interestingly, among all 25 × 4 runs, *CCGSLS* achieved the lowest error of 1.036E − 09.

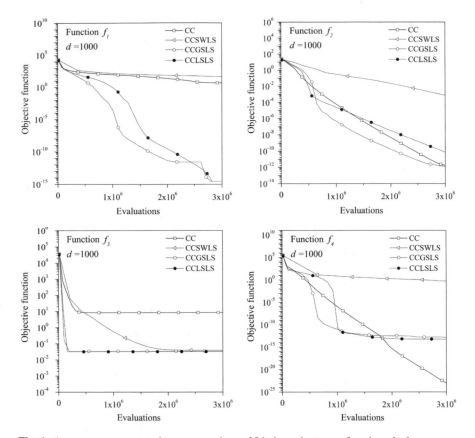

Fig. 1 Average convergence plots computed over 25 independent runs: functions f_1–f_4

The optimization of test function f_6 led to a negligible error for all algorithms except *CCSWLS*. According to the convergence plot in Fig. 2, the surrogate-assisted algorithms provided a very satisfactory error in only 5E + 05 fitness evaluations (i.e., around 10^{-8} and 10^{-9} for *CCGSLS* and *CCLSLS* respectively).

The shifted Rosenbrock function f_7 is well known for leading to hard optimization problems. The statistically equivalent best results were achieved by the surrogate-assisted algorithms, which led to the average errors of 8.622 and 9.104, for *CCGSLS* and *CCLSLS* respectively. However, among all runs the lowest error of 7.772E − 01 was achieved by the standard CC.

For function f_8, the algorithm *CCLSLS* largely outperformed all other algorithms by providing an average error of 4.130E − 15 with a very small standard deviation. However, the lowest errors provided by *CC*, *CCGSLS* and *CCLSLS* in the 25 independent runs were all negligible for practical purposes. The large difference in the average error between *CCGSLS* and *CCLSLS* when optimizing f_8 deserves a thorough investigation.

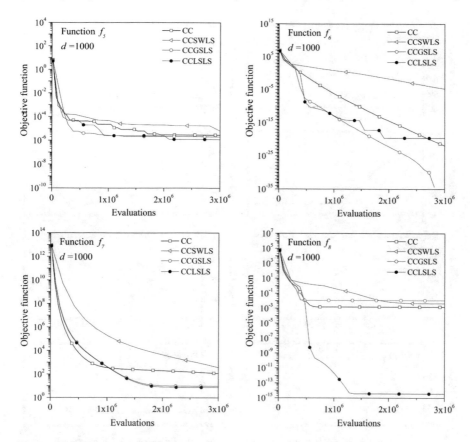

Fig. 2 Average convergence plots computed over 25 independent runs: functions f_5–f_8

Although a statistically sound comparison between algorithms would require a richer suite of test functions, the above preliminary results showed that in most relevant cases the surrogate-assisted memetic approach can make the search more effective and efficient. Moreover, in none of the examined optimization problems the surrogate approach to LS led to a significant worsening of final error.

Nevertheless, many configuration details of the proposed approach can significantly influence the results. As an example, Fig. 3 shows, for the *CCGSLS* algorithm, how the achieved average error varies with the number q of individuals whose neighbourhoods are the objects of LS in each subcomponent. As can be seen, for five functions (i.e., $f_1, f_2, f_3, f_6,$ and f_8) the best average result was obtained using $q = 15$. However, for function f_5 a much better result can be achieved by increasing the value of q. In contrast, for f_7 the lowest error corresponds to a single LS per subcomponent, initialized with the best individual of the population. By comparing Figs. 2 and 3, it is clear that in some case a suitable way to select the individuals for LS initialization can be crucial.

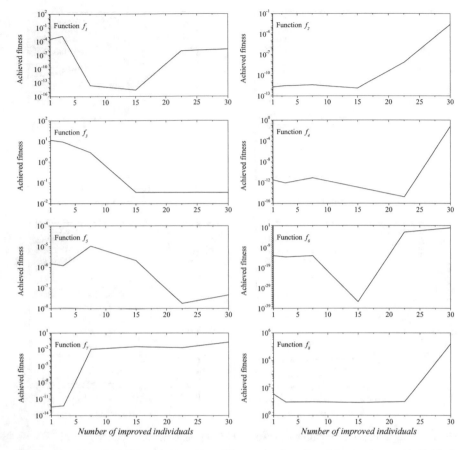

Fig. 3 Algorithm *CCGSLS*: average achieved fitness as a function of the number q of individuals improved with LS at each cycle and in each subcomponent

Although not reported here, also other settings of the surrogate-assisted memetic algorithm can play a significant role. For example, the intensity λ or the empirical rule for initializing and adapting the size of trust-region. Also the GP learning phase can be improved, especially regarding the selection of the training set and the exploitation of the model variance expressed by Eq. (12).

5   Directions of Further Research

The optimization of expensive and high-dimensional functions has attracted so far little research. The approach discussed in this chapter, in which fitness surrogates are built and exploited within low-dimensional subcomponents, defines a promising research direction. However, to devise an efficient optimization tool many aspects should be investigated and refined.

First, it is known that the size of subcomponents can strongly influence the search efficiency of a CC optimizer [59, 84]. However, in this case it also affects the quality of surrogates. Thus, determining a suitable decomposition, for example within an adaptive strategy, can be a critical factor of success for the optimization.

Moreover, it was shown above that the selection of individuals to be improved by LS is another important aspect. As mentioned in Sect. 2.2, in MA-LS-Chain only one individual is chosen at each generation. However, for most of the functions used with *CCGSLS* a greater number of improved individuals led to better results.

Also the role of LS intensity, which was kept at a very low fixed value in the above experimentation, should be the object of further study. In general, adapting the participation of LS and global search in consuming the available computational budget, as done for example in the MOS approach (see Sect. 2.2), could be a fruitful approach.

Another aspect that deserves to be studied concerns the use of more complex covariance functions, for example by considering an anisotropic matrix Λ of length scales.

Finally, the preliminary experiments presented above concern the enhancement of LS at the subcomponent level using surrogates. However, as illustrated in Sect. 2.3 there are several ways to exploit surrogates during the search. Future work should also address different uses of approximate fitness, also exploiting the GP posterior variance given by Eq. (12), as well as different ways of selecting the training set.

References

1. Abboud, K., Schoenauer, M.: Surrogate deterministic mutation: preliminary results. In: Collet, P., Fonlupt, C., Hao, J.K., Lutton, E., Schoenauer, M. (eds.) Artificial Evolution. Lecture Notes in Computer Science, vol. 2310, pp. 104–116. Springer, Berlin (2002)
2. Aguilar-Ruiz, J., Mateos, D., Rodriguez, D.: Evolutionary neuroestimation of fitness functions. In: Lecture Notes on Artificial Inteligence, vol. 2902, pp. 74–83 (2003)
3. Anderson, K., Hsu, Y.: Genetic crossover strategy using an approximation concept. In: Proceedings of the 1999 Congress on Evolutionary Computation, 1999. CEC 99. vol. 1, p. 533 (1999)
4. Bellman, R.: Dynamic Programming, 1st edn. Princeton University Press, Princeton (1957)
5. Bischl, B., Mersmann, O., Trautmann, H., Weihs, C.: Resampling methods for meta-model validation with recommendations for evolutionary computation. Evol. Comput. **20**(2), 249–275 (2012)
6. Blum, M., Riedmiller, M.A.: Optimization of Gaussian process hyperparameters using rprop. In: 21st European Symposium on Artificial Neural Networks, ESANN 2013, Bruges, Belgium, 24–26 April, 2013. https://www.elen.ucl.ac.be/esann/proceedings/papers.php?ann=2013
7. Branke, J., Schmidt, C.: Fast convergence by means of fitness estimation. Soft Comput. J. (2003, in press)
8. Brest, J., Maucec, M.S.: Self-adaptive differential evolution algorithm using population size reduction and three strategies. Soft Comput. **15**(11), 2157–2174 (2011)

9. Bueche, D., Schraudolph, N., Koumoutsakos, P.: Accelerating evolutionary algorithms with Gaussian process fitness function models. IEEE Trans. Syst. Man, Cybern.: Part C **35**(2), 183–194 (2004)
10. Carpenter, W., Barthelemy, J.F.: A comparison of polynomial approximation and artificial neural nets as response surface. Technical Report, 92–2247, AIAA (1992)
11. Celis, M., Dennis Jr., J., Tapia, R.: A trust region strategy for nonlinear equality constrained optimization. In: Proceedings of the SIAM Conference on Numerical Optimization, Boulder, CO (1984)
12. Chen, W., Weise, T., Yang, Z., Tang, K.: Large-scale global optimization using cooperative coevolution with variable interaction learning. In: Parallel Problem Solving from Nature. PPSN XI, Lecture Notes in Computer Science, vol. 6239, pp. 300–309. Springer, Berlin (2010)
13. Cheng, S., Ting, T., Yang, X.S.: Large-scale global optimization via swarm intelligence. In: Koziel, S., Leifsson, L., Yang, X.S. (eds.) Solving Computationally Expensive Engineering Problems. Springer Proceedings in Mathematics and Statistics, vol. 97, pp. 241–253. Springer, Berlin (2014)
14. Cheng, R., Jin, Y., Narukawa, K., Sendhoff, B.: A multiobjective evolutionary algorithm using Gaussian process based inverse modeling. IEEE Trans. Evol. Comput. **PP**(99), 1–1 (2015)
15. D'Ambrosio, D., Rongo, R., Spataro, W., Trunfio, G.A.: Meta-model assisted evolutionary optimization of cellular automata: an application to the sciara model. In: Parallel Processing and Applied Mathematics. Lecture Notes in Computer Science, vol. 7204, pp. 533–542. Springer, Berlin (2012)
16. D'Ambrosio, D., Rongo, R., Spataro, W., Trunfio, G.A.: Optimizing cellular automata through a meta-model assisted memetic algorithm. In: Parallel Problem Solving from Nature—PPSN XII. LNCS, vol. 7492, pp. 317–326. Springer, Berlin (2012)
17. Doerner, K., Hartl, R.F., Reimann, M.: Cooperative ant colonies for optimizing resource allocation in transportation. In: Proceedings of the EvoWorkshops on Applications of Evolutionary Computing, pp. 70–79. Springer, Berlin (2001)
18. El-Abd, M., Kamel, M.S.: A Taxonomy of cooperative particle swarm optimizers. Int. J. Comput. Intell. Res. **4** (2008)
19. El-Beltagy, M., Keane, A.: Evolutionary optimization for computationally expensive problems using Gaussian processes. In: Proceedings of International Conference on Artificial Intelligence, pp. 708–714. CSREA (2001)
20. Emmerich, M., Giotis, A., Özdenir, M., Bäck, T., Giannakoglou, K.: Metamodel-assisted evolution strategies. In: Parallel Problem Solving from Nature. Lecture Notes in Computer Science, vol. 2439, pp. 371–380. Springer, Berlin (2002)
21. Filho, F., Gomide, F.: Fuzzy clustering in fitness estimation models for genetic algorithms and applications. In: 2006 IEEE International Conference on Fuzzy Systems, pp. 1388–1395 (2006)
22. Giunta, A., Watson, L.: A comparison of approximation modeling techniques: polynomial versus interpolating models. Technical Report 98–4758, AIAA (1998)
23. Goh, C., Lim, D., Ma, L., Ong, Y., Dutta, P.: A surrogate-assisted memetic co-evolutionary algorithm for expensive constrained optimization problems. In: 2011 IEEE Congress on Evolutionary Computation (CEC), pp. 744–749 (2011)
24. Hansen, N., Ostermeier, A.: Completely derandomized self-adaptation in evolution strategies. Evol. Comput. **9**(2), 159–195 (2001)
25. Hasanzadeh, M., Meybodi, M., Ebadzadeh, M.: Adaptive cooperative particle swarm optimizer. Appl. Intell. **39**(2), 397–420 (2013)
26. Hong, Y.S., Lee, H., Tahk, M.J.: Acceleration of the convergence speed of evolutionary algorithms using multi-layer neural networks. Eng. Optim. **35**(1), 91–102 (2003)
27. Hüscken, M., Jin, Y., Sendhoff, B.: Structure optimization of neural networks for aerodynamic optimization. Soft Comput. J. **9**(1), 21–28 (2005)
28. Jin, Y.: A comprehensive survey of fitness approximation in evolutionary computation. Soft Comput. **9**(1), 3–12 (2005)

29. Jin, Y.: A comprehensive survey of fitness approximation in evolutionary computation. Soft Comput. J. **9**(1), 3–12 (2005)
30. Jin, Y.: Surrogate-assisted evolutionary computation: recent advances and future challenges. Swarm Evol. Comput. **1**(2), 61–70 (2011)
31. Jin, Y., Olhofer, M., Sendhoff, B.: On evolutionary optimization with approximate fitness functions. In: Proceedings of the Genetic and Evolutionary Computation Conference, pp. 786–792. Morgan Kaufmann (2000)
32. Jin, Y., Olhofer, M., Sendhoff, B.: A framework for evolutionary optimization with approximate fitness functions. IEEE Trans. Evol. Comput. **6**(5), 481–494 (2002)
33. Jin, Y., Huesken, M., Sendhoff, B.: Quality measures for approximate models in evolutionary computation. In: Proceedings of GECCO Workshops: Workshop on Adaptation, Learning and Approximation in Evolutionary Computation, pp. 170–174. Chicago (2003)
34. Jin, Y., Sendhoff, B.: Reducing fitness evaluations using clustering techniques and neural networks ensembles. In: Genetic and Evolutionary Computation Conference. LNCS, vol. 3102, pp. 688–699. Springer, Berlin (2004)
35. Kim, H.S., Cho, S.B.: An efficient genetic algorithm with less fitness evaluation by clustering. In: Proceedings of the 2001 Congress on Evolutionary Computation, 2001, vol. 2, pp. 887–894 (2001)
36. King, D.E.: Dlib-ml: a machine learning toolkit. J. Mach. Learn. Res. **10**, 1755–1758 (2009)
37. Lastra, M., Molina, D., Bentez, J.M.: A high performance memetic algorithm for extremely high-dimensional problems. Inf. Sci. **293**, 35–58 (2015)
38. LaTorre, A.: A framework for hybrid dynamic evolutionary algorithms: multiple offspring sampling (MOS). Ph.D. thesis, Universidad Politecnica de Madrid (2009)
39. LaTorre, A., Muelas, S., Peña, J.M.: A MOS-based dynamic memetic differential evolution algorithm for continuous optimization: a scalability test. Soft Comput. **15**(11), 2187–2199 (2011)
40. LaTorre, A., Muelas, S., Pena, J.M.: Multiple offspring sampling in large scale global optimization. In: 2012 IEEE Congress on Evolutionary Computation (CEC), pp. 1–8 (2012)
41. LaTorre, A., Muelas, S., Pena, J.M.: Large scale global optimization: experimental results with MOS-based hybrid algorithms. In: IEEE Congress on Evolutionary Computation (CEC), 2013, pp. 2742–2749 (2013)
42. LaTorre, A., Muelas, S., Peña, J.M.: A comprehensive comparison of large scale global optimizers. Inf. Sci. **316**, 517–549 (2015)
43. Li, X., Yao, X.: Cooperatively coevolving particle swarms for large scale optimization. IEEE Trans. Evol. Comput. **16**(2), 210–224 (2012)
44. Liu, B., Zhang, Q., Gielen, G.: A Gaussian process surrogate model assisted evolutionary algorithm for medium scale expensive optimization problems. IEEE Trans. Evol. Comput. **18**(2), 180–192 (2014)
45. Liu, J., Tang, K.: Scaling up covariance matrix adaptation evolution strategy using cooperative coevolution. In: Yin, H., Tang, K., Gao, Y., Klawonn, F., Lee, M., Weise, T., Li, B., Yao, X. (eds.) Intelligent Data Engineering and Automated Learning—IDEAL 2013. Lecture Notes in Computer Science, vol. 8206, pp. 350–357. Springer, Berlin (2013)
46. Liu, Y., Yao, X., Zhao, Q.: Scaling up fast evolutionary programming with cooperative coevolution. In: Proceedings of the 2001 Congress on Evolutionary Computation, Seoul, Korea, pp. 1101–1108 (2001)
47. Liu, W., Zhang, Q., Tsang, E., Virginas, B.: Fuzzy clustering based Gaussian process model for large training set and its application in expensive evolutionary optimization. In: IEEE Congress on Evolutionary Computation, 2009. CEC '09, pp. 2411–2415 (2009)
48. Lozano, M., Molina, D., Herrera, F.: Editorial: scalability of evolutionary algorithms and other metaheuristics for large-scale continuous optimization problems. Soft Comput. **15**(11), 2085–2087 (2011)
49. MacKay, D.J.C.: Introduction to Gaussian processes. In: Bishop, C.M. (ed.) Neural Networks and Machine Learning, NATO ASI Series, pp. 133–166. Kluwer Academic Press (1998)

50. Mahdavi, S., Shiri, M.E., Rahnamayan, S.: Metaheuristics in large-scale global continues optimization: a survey. Inf. Sci. **295**, 407–428 (2015)
51. Martnez, S.Z., Coello Coello, C.A.: A memetic algorithm with non gradient-based local search assisted by a meta-model. In: Schaefer, R., Cotta, C., Koodziej, J., Rudolph, G. (eds.) Parallel Problem Solving from Nature, PPSN XI, Lecture Notes in Computer Science, vol. 6238, pp. 576–585. Springer, Berlin (2010)
52. Molina, D., Lozano, M., García-Martínez, C., Herrera, F.: Memetic algorithms for continuous optimisation based on local search chains. Evol. Comput. **18**(1), 27–63 (2010)
53. Molina, D., Lozano, M., Herrera, F.: MA-SW-Chains: memetic algorithm based on local search chains for large scale continuous global optimization. In: IEEE Congress on Evolutionary Computation (CEC), 2010, pp. 1–8 (2010)
54. Molina, D., Lozano, M., Sánchez, A.M., Herrera, F.: Memetic algorithms based on local search chains for large scale continuous optimisation problems: MA-SSW-Chains. Soft Comput. **15**(11), 2201–2220 (2011)
55. Moscato, P.: On evolution, search, optimization, genetic algorithms and martial arts: towards memetic algorithms. Technical Report. Caltech Concurrent Computation Program Report 826, Caltech, Pasadena, California (1989)
56. Moscato, P.: New ideas in optimization. chap. Memetic Algorithms: A Short Introduction, pp. 219–234. McGraw-Hill Ltd., UK, Maidenhead, UK, England (1999)
57. Omidvar, M.N., Li, X., Yao, X.: Cooperative co-evolution with delta grouping for large scale non-separable function optimization. In: IEEE Congress on Evolutionary Computation, pp. 1–8 (2010)
58. Omidvar, M.N., Li, X., Mei, Y., Yao, X.: Cooperative co-evolution with differential grouping for large scale optimization. IEEE Trans. Evol. Comput. **18**(3), 378–393 (2014)
59. Omidvar, M.N., Mei, Y., Li, X.: Effective decomposition of large-scale separable continuous functions for cooperative co-evolutionary algorithms. In: Proceedings of the IEEE Congress on Evolutionary Computatio. IEEE (2014)
60. Ong, Y.S., Keane, A.: Meta-Lamarckian learning in memetic algorithms. IEEE Trans. Evolut. Comput. **8**(2), 99–110 (2004)
61. Ong, Y., Keane, A., Nair, P.: Surrogate-assisted coevolutionary search. In: Proceedings of the 9th International Conference on Neural Information Processing, 2002. ICONIP '02, vol. 3, pp. 1140–1145 (2002)
62. Ong, Y.S., Zhou, Z., Lim, D.: Curse and blessing of uncertainty in evolutionary algorithm using approximation. In: IEEE Congress on Evolutionary Computation, 2006. CEC 2006, pp. 2928–2935 (2006)
63. Parsopoulos, K.E.: Parallel cooperative micro-particle swarm optimization: a master-slave model. Appl. Soft Comput. **12**(11), 3552–3579 (2012)
64. Peremezhney, N., Hines, E., Lapkin, A., Connaughton, C.: Combining Gaussian processes, mutual information and a genetic algorithm for multi-target optimization of expensive-to-evaluate functions. Eng. Optim. **46**(11), 1593–1607 (2014)
65. Potter, M.A., De Jong, K.A.: A cooperative coevolutionary approach to function optimization. In: Proceedings of the International Conference on Evolutionary Computation. The Third Conference on Parallel Problem Solving from Nature: Parallel Problem Solving from Nature, PPSN III, pp. 249–257. Springer (1994)
66. Potter, M.A., De Jong, K.A.: Cooperative coevolution: an architecture for evolving coadapted subcomponents. Evolut. Comput. **8**(1), 1–29 (2000)
67. Qin, A., Huang, V., Suganthan, P.: Differential evolution algorithm with strategy adaptation for global numerical optimization. IEEE Trans. Evolut. Comput. **13**(2), 398–417 (2009)
68. Rasheed, K., Hirsh, H.: Informed operators: speeding up genetic-algorithm-based design optimization using reduced models. In: Proceedings of the Genetic and Evolutionary Computation Conference (GECCO), pp. 628–635. Morgan Kaufmann (2000)
69. Rasmussen, C.E., Williams, C.K.I.: Gaussian Processes for Machine Learning (Adaptive Computation and Machine Learning). The MIT Press (2005)

70. Regis, R., Shoemaker, C.: Constrained global optimization of expensive black box functions using radial basis functions. J. Global Optim. **31**(1), 153–171 (2005)
71. Riedmiller, M., Braun, H.: A direct adaptive method for faster backpropagation learning: the rprop algorithm. In: IEEE International Conference on Neural Networks, pp. 586–591 (1993)
72. Schmitz, A., Besnard, E., Vivies, E.: Reducing the cost of computational fluid dynamics optimization using multilayer perceptrons. In: IEEE 2002 World Congress on Computational Intelligence. IEEE (2002)
73. Sheskin, D.J.: Handbook of Parametric and Nonparametric Statistical Procedures, 4tn edn. Chapman & Hall/CRC (2007)
74. Snchez-Ante, G., Ramos, F., Frausto, J.: Cooperative simulated annealing for path planning in multi-robot systems. In: MICAI 2000: Advances in Artificial Intelligence. LNCS, vol. 1793, pp. 148–157. Springer, Berlin (2000)
75. Solis, F.J., Wets, R.J.B.: Minimization by random search techniques. Math. Oper. Res. **6**(1), 19–30 (1981)
76. Sun, L., Yoshida, S., Cheng, X., Liang, Y.: A cooperative particle swarm optimizer with statistical variable interdependence learning. Inf. Sci. **186**(1), 20–39 (2012)
77. Tang, K., Yao, X., Suganthan, P., MacNish, C., Chen, Y., Chen, C., Yang, Z.: Benchmark functions for the CEC' 2008 special session and competition on large scale global optimization
78. Tang, K., Li, X., Suganthan, P.N., Yang, Z., Weise, T.: Benchmark functions for the CEC'2010 special session and competition on large-scale global optimization. http://nical.ustc.edu.cn/cec10ss.php
79. Tang, K., Yang, Z., Weise, T.: Special session on evolutionary computation for large scale global optimization at 2012 IEEE world congress on computational intelligence (cec@wcci-2012). Technical report, Hefei, Anhui, China: University of Science and Technology of China (USTC), School of Computer Science and Technology, Nature Inspired Computation and Applications Laboratory (NICAL) (2012)
80. Tenne, Y., Armfield, S.: A Memetic Algorithm Assisted by an Adaptive Topology RBF Network and Variable Local Models for Expensive Optimization Problems. INTECH Open Access Publisher (2008)
81. Tenne, Y., Armfield, S.: Metamodel accuracy assessment in evolutionary optimization. In: IEEE Congress on Evolutionary Computation, 2008. CEC 2008. (IEEE World Congress on Computational Intelligence), pp. 1505–1512 (2008)
82. Tenne, Y., Armfield, S.: A framework for memetic optimization using variable global and local surrogate models. Soft Comput. **13**(8–9), 781–793 (2009)
83. Trunfio, G.A.: Enhancing the firefly algorithm through a cooperative coevolutionary approach: an empirical study on benchmark optimisation problems. IJBIC **6**(2), 108–125 (2014)
84. Trunfio, G.A.: A cooperative coevolutionary differential evolution algorithm with adaptive subcomponents. Proc. Comput. Sci. **51**, 834–844 (2015)
85. Tseng, L.Y., Chen, C.: Multiple trajectory search for large scale global optimization. In: IEEE Congress on Evolutionary Computation, 2008. CEC 2008. (IEEE World Congress on Computational Intelligence), pp. 3052–3059 (2008)
86. Ulmer, H., Streichert, F., Zell, A.: Evolution strategies assisted by Gaussian processes with improved pre-selection criterion. In: Proceedings of IEEE Congress on Evolutionary Computation, pp. 692–699 (2003)
87. Van den Bergh, F., Engelbrecht, A.P.: A cooperative approach to particle swarm optimization. IEEE Trans. Evol. Comput. **8**(3), 225–239 (2004)
88. Wang, Y., Huang, J., Dong, W.S., Yan, J.C., Tian, C.H., Li, M., Mo, W.T.: Two-stage based ensemble optimization framework for large-scale global optimization. Eur. J. Oper. Res. **228**(2), 308–320 (2013)
89. Wang, Y., Li, B.: Two-stage based ensemble optimization for large-scale global optimization. In: IEEE Congress on Evolutionary Computation (CEC), 2010, pp. 1–8 (2010)
90. Willmes, L., Baeck, T., Jin, Y., Sendhoff, B.: Comparing neural networks and kriging for fitness approximation in evolutionary optimization. In: Proceedings of IEEE Congress on Evolutionary Computation, pp. 663–670 (2003)

91. Won, K., Ray, T., Tai, K.: A framework for optimization using approximate functions. In: Proceedings of IEEE Congress on Evolutionary Computation, pp. 1077–1084 (2003)
92. Yang, Z., Tang, K., Yao, X.: Large scale evolutionary optimization using cooperative coevolution. Inf. Sci. **178**(15), 2985–2999 (2008)
93. Yang, Z., Tang, K., Yao, X.: Multilevel cooperative coevolution for large scale optimization. In: IEEE Congress on Evolutionary Computation, pp. 1663–1670. IEEE (2008)
94. Yang, Z., Tang, K., Yao, X.: Self-adaptive differential evolution with neighborhood search. In: IEEE Congress on Evolutionary Computation, 2008. CEC 2008. (IEEE World Congress on Computational Intelligence), pp. 1110–1116 (2008)
95. Yang, Z., Tang, K., Yao, X.: Scalability of generalized adaptive differential evolution for large-scale continuous optimization. Soft Comput. **15**(11), 2141–2155 (2011)
96. Zhang, J., Sanderson, A.: Jade: Adaptive differential evolution with optional external archive. IEEE Trans. Evol. Comput. **13**(5), 945–958 (2009)
97. Zhang, Q., Liu, W., Tsang, E., Virginas, B.: Expensive multiobjective optimization by moea/d with Gaussian process model. IEEE Trans. Evol. Comput. **14**(3), 456–474 (2010)
98. Zhou, Z., Ong, Y., Lim, M., Lee, B.: Memetic algorithm using multi-surrogates for computationally expensive optimization problems. Soft Comput. **11**(10), 957–971 (2007)
99. Zhou, Z., Ong, Y.S., Nair, P.B., Keane, A.J., Lum, K.Y.: Combining global and local surrogate models to accelerate evolutionary optimization. IEEE Trans. Syst. Man Cybern. Part C **37**(1), 66–76 (2007)
100. Zhou, Z., Ong, Y.S., Nguyen, M.H., Lim, D.: A study on polynomial regression and Gaussian process global surrogate model in hierarchical surrogate-assisted evolutionary algorithm. In: The 2005 IEEE Congress on Evolutionary Computation, 2005, vol. 3, pp. 2832–2839 (2005)

Cuckoo Search: From Cuckoo Reproduction Strategy to Combinatorial Optimization

Aziz Ouaarab and Xin-She Yang

Abstract Combinatorial optimization problems, specially those that are NP-hard, are increasingly being dealt with by stochastic, metaheuristic approaches. Most recently developed metaheuristics are nature-inspired and they are often inspired by some special characteristics in evolution, ecological or biological systems. This chapter discusses how to go from a biological phenomenon such as the aggressive reproduction strategy of cuckoos to solve tough problems in the combinatorial search space. Key features and steps are highlighted, together with the discussions of further research topics.

1 Introduction

Many combinatorial optimization problems are non-deterministic polynomial-time hard (or NP-hard) and there are no efficient algorithms to solve such hard optimization problems. That is to say that there is no algorithm that can find its optimal solution in a polynomial time on a deterministic machine [10]. Thus, there is a strong need to use alternative methods so as to find a good quality solution in a practically acceptable timescale and these alternative approaches tend to be approximate or stochastic algorithms such as nature inspired metaheuristics [3, 13].

On the other hand, the way and knowledge we know about the problem and any problem-specific knowledge will be very useful to guide the search process. But for an algorithm to be more general, sometimes, a black-box approach may also be desired because this may provides some flexibility. Thus, the problem can be viewed

A. Ouaarab (✉)
LRIT, Associated Unit to the CNRST (URAC 29), Mohammed V-Agdal
University, B.P. 1014 Rabat, Morocco
e-mail: aziz.ouaarab@gmail.com

X.-S. Yang
School of Science and Technology, Middlesex University,
The Burroughs, London NW4 4BT, UK
e-mail: x.yang@mdx.ac.uk

© Springer International Publishing Switzerland 2016
X.-S. Yang (ed.), *Nature-Inspired Computation in Engineering*,
Studies in Computational Intelligence 637, DOI 10.1007/978-3-319-30235-5_5

from a new alternative perspective and such new type of designs of the problem may allow more flexibility for searching in the solution space.

Nature-inspired computing tends to mimic an autonomous behaviour of individuals in a multi-agent system that act independently and follow simple local rules, and the key of most nature-inspired optimization designs is the interaction of these individuals in a swarm or population by considering their adaptation techniques to the dynamically changing environment. This requires a good understanding of complex nature phenomena and ecological behaviours.

Natural systems are usually complex and thus not easy to understand. The seemingly stochastic behaviour of natural systems may appear unpredictable, however, such ecosystems can show a substantial degree of balance, adaptivity and optimality in an interacting ecological system [4] and a good example is the obligate interspecific brood parasitism [34] such as those of cuckoos [35]. Based on this aggressive reproduction strategy, the cuckoo search (CS) algorithm was developed by Xin-She Yang and Suash Deb in 2009 [49], and this method has been extended to solve combinatorial optimization problems [6, 17]. Most of this cuckoo theme follows a common model that allows the ease to pass from natural system to the main search components in a combinatorial space.

The present chapter will discuss the common model and search approaches and describe their main three components: solution, neighbourhood, and moves. The remainder of this chapter is organized as follows. Section 2 introduces the inspiration concept which combines cuckoos' brood parasitism with Lévy flights. Section 3 presents metaheuristics, their definition and some related notions. Section 4 gives some statements and rules of the cuckoo search approach. Section 5 first briefly describes the combinatorial space, then discusses the search space model, and cites some constraints of the whole process. Section 6 gives a selection of applications of the cuckoo search approach to some combinatorial optimization problems. Finally, Sect. 7 concludes with some discussions.

2   Inspiration from Nature

Nature has always been an important source of inspiration for engineering design and optimization. In fact, natural systems can be considered as the first optimizers because living beings in nature evolve and adapt their behaviour according to the changing environment; otherwise, they may extinct, or at least the whole ecosystem may have to be reorganized to form a different evolutionary equilibrium.

An ecosystem as a community of living interacting organisms that interact and respond to their non-living environment forms an autonomous and distributed system in a heterogeneous, dynamically varying environment. The interactions and complex relationships among different organisms may lead to diversity and richness of species in this environment. Such interactions can be summarized as intra- and inter-specific competition for abiotic and biotic resources, predation, parasitism and mutualism [16].

Some multiple organisms in the same species such as birds and fish may form a swarm under certain conditions, while many can form a complex interacting system even with different species. Among many fascinating behaviours in nature, the so-called cuckoo brood parasitism is one of the pertinent examples of an evolved ecological behaviour. For the cuckoo parasitism, some cuckoo species use other birds species such as warblers to rear their young, which is essentially a distributed intelligent reproduction strategy. This aggressive strategy combined with Lévy Flights distribution forms a source of inspiration for the cuckoo search algorithm.

2.1 Brood Parasitism

Cuckoo are fascinating birds, not only because of the beautiful sounds they can make, but also because of their aggressive reproduction strategy. Some species such as the *ani* and *Guira* cuckoos lay their eggs in communal nests, though they may remove others' eggs to increase the hatching probability of their own eggs. Quite a number of species engage the obligate brood parasitism by laying their eggs in the nests of other host birds (often other species).

There are three basic types of brood parasitism: intraspecific brood parasitism, cooperative breeding, and nest takeover. Some host birds can engage direct conflict with the intruding cuckoos. If a host bird discovers the eggs are not their owns, they will either get rid of these alien eggs or simply abandon its nest and build a new nest elsewhere. Some cuckoo species such as the New World brood-parasitic *Tapera* have evolved in such a way that female parasitic cuckoos are often very specialized in the mimicry in colour and pattern of the eggs of a few chosen host species. This reduces the probability of their eggs being abandoned and thus increases their reproductivity.

In addition, the best timing of egg-laying of some species is that cuckoos tends to lay in a host where host eggs have just been laid but just before the incubation starts. Once the first cuckoo chick is hatched, the first instinct action it will take is to evict the host eggs by blindly propelling the eggs out of the nest, which increases the cuckoo chick's share of food provided by its host bird. Studies also show that a cuckoo chick can also mimic the call of host chicks to gain access to more feeding opportunity [43].

It can be expected that such parasitic behaviour can have evolutionary advantages so that it saves time and efforts in best building and rearing young cuckoo chicks, and thus cuckoos can focus efforts of maximizing egg-laying in multiple host nests so as to increase their reproductive capability.

2.1.1 Swarm Intelligence and Evolution

Many biological agents such as insects, ants, bees and fish may show some swarming behaviour, and such swarming characteristics may often emerge from a complex system with multiple, interacting agents that follow some seemingly simple and local rules. These rules are decentralized and there is no single leader in the system. Under certain conditions, such local rules can lead to some self-organization and higher-level characteristics that are almost impossible to achieve by seemingly unintelligent individuals [11, 18].

For example, particle swarm optimization was developed based on the swarming characteristics of fish schooling and bird swarming [18]. Another example is the cuckoo search that was developed based on cuckoo behaviour.

During the evolution process, the genetic traits of a biological population will evolve and change, leading to diversity and adaptation of species [15]. In case of cuckoo species, some cuckoo species have evolved to develop effort-saving strategy by using brooding parasitism and taking advantages of the host birds such as warblers. In order to increase the probability of cuckoos' success, cuckoos can lay eggs that are sufficiently similar to the eggs of the host birds in terms of colours and texture. At the same time, host bird species also evolve to increase their abilities to recognize the eggs laid by cuckoos. This arm race between cuckoos and host birds has been evolving over millions of years [35], thus forming a kind of co-evolution [8]. Obviously, such co-evolution provides a new perspective for developing new competitive nature-inspired algorithms and extending the current cuckoo search strategies.

2.2 Lévy Flights

Many local moves in the biological systems may be governed by the local random walks. However, some animals when searching for food sources may occasionally venture much further, leading to a quasi-random search manner. In fact, quite a few studies suggested that some animals and birds such as albatross carry out Lévy Flights when searching for preys. This Lévy flights can be commonly represented by a random search around the current solution and occasional big jumps, which be approximated by a power law [5, 45]

$$N(s) \sim \frac{\lambda \Gamma(\lambda) \sin(\pi \lambda / 2)}{\pi |s|^{1+\lambda}}, \tag{1}$$

where step sizes s are sufficiently large. Here, the notation $\sim$ means that $N(s)$ is drawn from the distribution on the right-hand side. In addition, $0 \leq \lambda \leq 2$ is the Lévy exponent and $\Gamma(\lambda)$ is the standard Gamma function

$$\Gamma(n) = \int_0^\infty t^{n-1} e^{-t} dt. \tag{2}$$

Lévy flights named after the French mathematician Paul Lévy [2, 40]. Studies show that Lévy flights can maximize the efficiency of resource searches in uncertain environments. In fact, Lévy flights have been observed among foraging patterns of albatrosses and fruit flies, and spider monkeys. Even humans such as the Ju/hoansi hunter-gatherers can trace paths of Lévy-flight patterns. In addition, Lévy flights have many applications. Many physical phenomena such as the diffusion of fluorescent molecules, cooling behavior and noise could show Lévy-flight characteristics under the right conditions [47].

When the characteristics of cuckoo's brood-parasitism behaviour is combined with Lévy flights, it becomes an efficient approach to solve optimization problems [48]. Lévy flights can naturally play an important role with local search and global search capabilities, which can simultaneously carry out search around the best solution and global search towards unexplored regions in the solution space.

3 Metaheuristics

The main avenue to pass from inspiration in nature to solving optimization problems is the so-called metaheuristics. To design a good metaheuristic algorithm, it often requires a good understanding of how the main successful characteristics in natural or biological systems work and then find mathematical links so that an appropriate search action can be achieved in the search space (in the present case, the combinatorial search space).

Metaheuristic algorithms usually use strategic search to effectively explore the solution space, and often focus on promising areas. Typically, these methods start with a set of initial solutions or initial sampling points in the search space, and then new solutions are generated based on the current population and the transition probability or moves in an iterative manner. This procedure continues until a predefined stopping criterion is met. Metaheuristics can have some advantages over traditional algorithms such as the Newton-Raphson method because explorative moves are often used in metaheuristics to explore unsearched region more effectively, and thus it is more likely to find the global optimal solutions to a given type of problem. However, metaheuristics may also have some disadvantages as well because they tend to require more computational efforts because their reduced ability in exploiting problem-specific information such as gradients.

However, for tough optimization problems in engineering and many disciplines, traditional methods struggle to cope, and metaheurisitc algorithms provide a promising set of alternative approaches. In addition, metaheuristics have simplicity and flexibility because they are often simple to implement, and yet they can solve many real-world optimization problems, from engineering optimization to applications concerning operations research, artificial intelligence and computational intelligence.

Among the tough optimization problems, a large part of combinatorial optimization problems may be related to NP-hard problems. At the moment, there is no efficient algorithm that can deal with this type of problems effectively. Thus, there is a strong need to quickly find a good solution (not necessarily the best) to these problems and thus it necessitates to develop different approximate algorithms such as relaxation methods and metaheuristics [41]. From a quick look at the recent studies and the expanding literature, we can see that metaheuristics have proven their efficiency and potential to solve a wide range of optimization problems. All this may be attracted to flexibility and versatility in dealing with different types of problems with the diversity of objective landscapes, whether continuous, discrete or mixed.

Almost all metaheuristic algorithms are population based and most are inspired by the collective behaviour of groups, colonies or swarms of some biological species. Such collective behaviours are somehow mimicked and adopted by artificial swarms consisting multiple agents or particles in order to find a source of food, or to search a partner, or to avoid predators over a vast foraging or search area. Tasks are achieved by distributed actions using relatively simple local rules with some information sharing ability.

On the other hand, from the algorithm analysis point of view, there are two key components in almost all metaheuristic algorithms and they are: intensification and diversification or exploitation and exploration. Diversification carries out explorative moves that usually in new regions in the solution space, while intensification typically exploit the information gathered during the search process to focus on the regions found so far that may be promising for better solutions in their neighbourhoods. In some literature, though exploitation and exploration are equivalent to intensification and diversification, some did argue that exploitation and exploration may designate short-term strategies with some degree of randomization, while intensification and diversification may be for medium and long term strategies [3, 23]. However, such differences may be subjective and may depend on the perspective of the analysis. In the one sense, intensification may use historically good solutions and/or memory, such search are mainly local around some elite solutions found so far in the previous search process [37]. Indeed, the exploitation of relevant information gathered from the sampling of the objective landscape may be crucial to the speedup of the convergence to the good or best solutions. Such information can also be useful to guide search to avoid local optima, though the exact role or avenue to achieve this may not be so clear.

In essence, diversification can guide the search process to explore unexplored areas in order to detect any promising regions where it may potentially contain the true global optimality. Ideally, such diversification should be able to visit a number of different regions in a more global scale in the solution space [3, 47]. However, it is not always easy to achieve it. As diversification requires many iterative explorative moves, it usually can slow down the search process with a slower convergence rate, but this will increase the probability of finding the true global optimal solutions. Therefore, a natural question is how to balance intensification and diversification? This is still an open question without a satisfactory answer at the moment. Generally speaking, the balance intensification and diversification may not be achieved easily and their role will depend on the algorithm, though such a balance should be dynamic and varying if it exists. Both intensification and diversification may require techniques subtly in terms of operators, randomization, actions and various strategies within a metaheuristic [46].

In addition, the classification of metaheuristics is not an easy task, though they can be grouped into several categories, based on different perspectives such as the source of inspiration, number of solutions or populations, and number of objective functions and others [3]. In the following section, we will give some descriptions about cuckoo search algorithm, and we will discuss how can we extend it to solve combinatorial optimization problems.

4   Cuckoo Search Approach

Inspired by the aggressive reproduction strategy of some cuckoo species, cuckoo search (CS) [48] uses the following ideal rules:

- Each cuckoo lays an egg at a time in a randomly chosen nest;
- A good quality egg in a nest can pass to a new generation;
- The number of host nests is fixed, and the egg laid by a cuckoo can be discovered by the host bird with a probability $p_a \in [0, 1]$.

With these idealized rules, it is necessary to establish some relationships between the source of inspiration and the search space (continuous or combinatorial spaces), but first the important statements or interpretations can be as follows:

- Each cuckoo egg in a nest represents a solution and each nest contains one egg. Though there may be advantageous for a cuckoo to lay multiple eggs and each nest should have more than one egg, the simplicity for one egg in one nest is purely for the ease of implementation here.
- Each cuckoo bird will lay a single egg at a time, and will choose its nest randomly. So, each individual in the population of cuckoos has the right to randomly generate a single new solution. This randomization is mainly diversification.
- The best nests with good quality eggs will be passed onto the new generation. Here, we implicitly introduced the concept of intensification or search around the best solutions.
- Some new solutions must be generated via Lévy flights around the best current solution. This step is both intensification (around the best solution) and diversification via Lévy flights.
- The number of host nests is fixed, and egg laid by the bird is discovered by the host with a probability $P_a \in [0, 1]$. In this case, the host bird can either choose to get rid of the egg, or abandon the nest and rebuild another nest elsewhere. For simplification, this hypothesis is approximated by the fraction p_a of abandoned nests among n nests. Again, this step is diversification.
- A large fraction of new solutions have to be generated randomly towards distant regions and their locations must be far enough from the current best solution, which will make the system not trapped in a local optimum. This is strong exploration and diversification.
- In the case that each nest may contain multiple eggs, it represents a set of solutions.

Algorithm 1 Cuckoo Search

1: Objective function $f(x)$, $x = (x_1, \ldots, x_d)^T$
2: Generate initial population of n host nests $x_i (= 1, 2, \ldots, n)$
3: **while** (t <MaxGeneration) or (stop criterion) **do**
4:    Get a cuckoo randomly by Lévy flights
5:    Evaluate its quality/fitness F_i
6:    Choose a nest among n (say, j) randomly
7:    **if** ($F_i > F_j$) **then**
8:       Replace j by the new solution i;
9:    **end if**
10:    A fraction (p_a) of worse nests are abandoned and new ones are built;
11:    Keep the best solutions (or nests with quality solutions);
12:    Rank the solutions and find the current best
13: **end while**
14: Postprocess results and visualization

The main step in the cuckoo search can be written as

$$X_i^{t+1} = X_i^t + \alpha s \otimes H(P_a - \epsilon) \otimes (X_j^t - X_k^t), \tag{3}$$

where X_j^t and X_k^t are two different solutions selected randomly by random permutation, $H(u)$ is a Heaviside function, ϵ is a random number drawn from a uniform distribution, and s is the step size drawn from a Lévy distribution. When generating new solutions $X^{(t+1)}$ on a global scale, a Lévy flight is performed

$$X_i^{(t+1)} = X_i^{(t)} + \alpha \oplus \text{Lévy}(\gamma), \tag{4}$$

where $\alpha > 0$ is the step size which should be related to the scales of the problem of interests. In most cases, we can use $\alpha = O(L/10)$ where L is the characteristic scale of the problem of interest. The above equation is essentially the stochastic equation for a random walk. In general, a random walk is a Markov chain whose next status/location only depends on the current location (the first term in the above equation) and the transition probability (the second term). The product $\oplus$ means entrywise multiplications. This entrywise product is similar to those used in PSO, but here the random walk via Lévy flight is more efficient in exploring the search space, as its step length is much longer in the long run.

In general, such search moves form random walks that again form Markov chains where the next step depends only on the current step. Mathematically speaking, random walks by Lévy flights are more effective in the exploration of the search space than other random walks because the sizes of the moves are typically much greater in the long term.

Recent studies show that the cuckoo search algorithm has two abilities: a local search and another global, controlled by the switch/probability of discovery parameter [27, 47]. Global search is favoured by P_a, while local search is strengthened in $1 - P_a$ of the population. This allows cuckoo search to effectively perform while managing a good balance between exploration and exploitation. Another advantageous factor is the choice of Lévy flights, instead of Gaussian random walks, which can carry out both local and global search simultaneously with a reduced number of parameters and often reduced computational times.

5 Combinatorial Space

Combinatorial optimization essentially concerns the analysis and solution of problems for the minimization or maximization of a value measure over a feasible space, typically involving mutually exclusive, logical constraints. In essence, such logical constraints can be considered as the arrangement of elements into some sets [33]. In reality, most combinatorial optimization problems (COP) arise from both theoretical problems and practical applications related to engineering design, computer science, transport planning and scheduling problems. Though the search space is discrete, the number of potential combinations of solutions in the feasible space can be astronomically large. The aim of any COP is to try to find a feasible solution from a (large) finite or infinitely countable set in a practically acceptable timescale. Such solutions can typically be an integer, subset, or a graph structure [32].

Though the exact formulation of a combinatorial optimization problem may vary, however, a particular formulation is not enough to determine the choice of solution methods, though it may affect the size of the feasible search space. As the number of combinations typically increases exponentially with the problem size, it is impossible to carry out any explicit enumeration, and thus simple approaches such as greedy techniques and dynamic programming can be extended to problems with NP-complexity. In this case, approximate methods and alternative methods such as nature-inspired metaheuristics may be desired [3, 13, 47]. Obviously, for both theoretical interest and practical importance, one of most studies problems in combinatorial optimization is the class of NP-hard problems [1], however, none of the existing algorithms are effective in dealing with such problems. But if the requirements are relaxed so as to overcome the complexity constraint, it is possible to seek good solutions intelligently that are not necessarily optimal, but the running can be carried out in an acceptable time.

In many aspects, nature can be considered as a truly first optimizer or even the best optimizer in problem solving. For example, ants and bees can often find the shortest routes in a given setting of the foraging surroundings. In the context of cuckoo search using the cuckoo reproduction strategy, often the nearest or the best nest can be chosen to lay eggs so as to increase the survival probability of the cuckoo chicks. In this case, the search or decision space is discrete because the numbers of nest choices and eggs laid are also integers.

In the next section, we will show how to transform cuckoo search with Lévy flights into a tool to solve combinatorial optimization problems.

5.1 *Modelling the Search Space*

A vast majority of combinatorial optimization problems can have a search space with feasible solutions represented as integers or binaries, including the assignment problems [20], knapsack problems [24], the travelling salesman problems [21], scheduling [36], vehicle routing [42] and other problems [26, 32, 38, 44]. Each solution, or sometimes a node, in the search space can be characterized by its position in terms of a combination of solution permutations, and each solution is also associated with a quality or fitness measure that be calculated from the objective function. In many cases, a combinatorial optimization problem can be represented by a graph, and all the neighbour solutions are connected by arcs. These arcs provides the means for the algorithmic operations that can change the position from the current solution to a new solution in the neighbourhood. Therefore, our search space can be typically represented by a graph as shown in Fig. 1.

The search space is usually encoded with a set of integer combinations. In the context of the cuckoo search, the set of all possible solutions may correspond to the combination set of spatial locations of the nests and the choice of which nests at what time. Consequently, moving between solutions is a change of positions from one nest to another of the same set or the set in the next generation. To provide more details

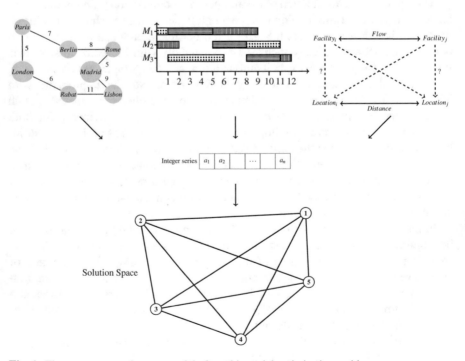

Fig. 1 The common search space model of combinatorial optimization problems.

about the proposed search space model, we can subdivide its main components as solutions, neighbourhoods and moves to be discussed in the rest of this section.

5.1.1 Solutions

The key component of the combinatorial search space is its solutions. However, the ways of representing solutions can be crucially important. A solution should be represent a certain combination of the values of all the decision variables and should be easy to validate its feasibility. In the case of cuckoo search, a solution corresponds to an egg and multiple solutions (eggs) can exist in a nest. However, for simplicity, it will be desirable that each nest only contains one egg, which means an egg is equivalent to a solution and a nest also corresponds to the same solution. In this case, an egg, a nest and a solution are essentially the same, and thus there is no need to distinguish them. An abandoned nest implies the replacement of an individual in the population by a new one. The discrete variables are the components of the solution that directly affect its quality or fitness value through the objective function of the problem, and thus this function must be well defined so that the representation of the coordinates are properly represented [28, 29].

5.1.2 Neighbourhood

Solutions in the combinatorial search space are connected by arcs. The one and only condition to connect a given pair of solutions, in the same space topology, is the so-called neighbourhood. The notion of a neighbourhood in the combinatorial space requires that the neighbour of a given solution has to be generated by the smallest possible perturbation. This perturbation must make the minimum changes on the current solution. It is therefore to specify subsets of solutions, namely the regions, according to the metric of the search space. In fact, such a neighbourhood notion is very important to effectively search good solutions in the combinatorial space so that the solutions are very similar to the existing solutions and not too far away.

As mentioned before, all nature-inspired metaheuristic approaches have to control and balance both intensification and diversification. Intensification focuses on the regions found so far and exploits the promising regions by searching in neighbourhoods of their best solutions, while diversification tends to explore new regions in the search space that might not be in the neighbourhood or even far away regions [23]. In order to achieve the balance between these two components, we have to differentiate neighbour solutions from distant or far away solutions.

5.1.3 Moves

In the combinatorial case, the solutions or solution coordinates in the search space are usually modified through characteristics of the problem under consideration.

Generally speaking, the positions in the combinatorial space can be changed by changing the order of a sequence where the solution components are manipulate via a combination, a permutation, or a set of methods or operators named perturbations or moves.

In many formulations, a step is the distance between two solutions, depending on the space topology and the neighbourhood notion. Steps are generally classified in terms of their lengths, nature of the proposed perturbations and its successive manipulation on the current solution carried out by a given algorithm.

Studies show that new solutions can be generated more effectively via Lévy flights. So, to enhance the search quality the step lengths are associated to values generated by Lévy flights [27, 47].

5.2 Constraints

Most design problems and their formulations are subject to constraints. To apply nature-inspired algorithms to discrete optimization problems, it is necessary to handle the encoding proposed solution to discrete dimensions and treat all constraints properly.

Most of such handling techniques to solve optimization problems are performed initially on continuous spaces using some relaxation techniques. Then, constraints and tight techniques are used to make better approximations. For example, the displacement in the continuous space from one solution to a neighbouring solution can be carried out by a variation or change in the coordinates or the order of a solution sequence. This perturbed solution produces in general a small difference between the qualities of the current solution and its neighbour. However, in the combinatorial space, even a minimum change to an existing solution may lead to a new solution that has remarkably different quality. This highlights the issue and possibility of finding the optimal solution and bad solutions in the same neighbourhood. Thus, care should be taken in using neighbourhood search. Therefore, it is worth pointing out that a small move from a current solution to its neighbourhood in the combinatorial space does not necessarily imply the small difference in solution quality.

6 Example of Applications

Metaheuristic algorithms have been applied to solve combinatorial optimization problems. The rest of this chapter will focus on the application of cuckoo search for solving various combinatorial problems and the comparison of the results with those in the literature. Other methods in the literature have obtained diversed results with different approaches in terms of solutions, neighbourhood and moves have also been used [14, 19]. Here we will provide a set of examples to show the diversity of the applications.

Table 1 Numerical results of the discrete cuckoo search for quadratic assignment problems

Instance	Bkv	Best	Average	Worst
tai12a	224416	**224416**	227769,6	230704
tai12b	39464925	**39464925**	**39464925**	**39464925**
tai15a	388214	**388214**	388643.06	389718
tai15b	51765268	**51765268**	**51765268**	**51765268**
tai20a	703482	705622	710536.9	717390
tai20b	122455319	**122455319**	122566157.8	123009513
tai25a	1167256	1171944	1181543,733	1191072
tai25b	344355646	**344355646**	344383574.2	344565108
tai30a	1818146	1825262	1838228,4	1849970
tai30b	637117113	**637117113**	641132438.2	651247567

6.1 Quadratic Assignment Problem

Quadratic assignment problems concern the assignment of n facilities to n locations, proportional to the flow between the facilities multiplied with their distances. So, the objective is to allocate each facility to each location in such a way that the total cost ($flow \times distance$) is minimized [20].

Dejam et al. [9] and Ouaarab et al. [31] tried to solve the quadratic assignment problem by considering the solution space as a set of all potential assignments of the facilities to the possible locations. This discrete version of cuckoo search defines a solution as a random permutation, where its size is the number of locations or facilities.

In the second approach, the operator used to move from one solution to a new one is 'swap operator', which represents either a small step or a big step, is generated by a series of consecutive swaps, and the size of this series is related to the values generated by Lévy flights, which allows us to have the choice to move in the neighbourhood by applying one small step or on the outside of the neighbourhood by a big step represented in this case by a series of small steps. Typical results obtained by the discrete cuckoo search are summarized in Table 1 where the bold font indicates that our results are better than the best results published in the literature.

6.2 Knapsack Problem

Knapsack problems concern a class of problems with diverse applications. In essence, for a given set of objects and each of the objects has a mass and a value, the aim is to pack the objects in a knapsack such that the total value is the maximum, subject to a fixed capacity of the knapsack. Therefore, the knapsack problem is a good example of combination optimization problems subject to the capacity constraint [24].

Gherboudj et al. [12] have applied a discrete version of the cuckoo search to transform a given solution from the continuous space to the discrete one as a binary solution by using the Sigmoid function in the interval [0, 1]. To obtain a binary solution, a random number is generated in the interval [0, 1] for each dimension i of the solution, and the generated numbers are compared with the flipping chance of the value of the Sigmoid function.

Local exploitation (moving in the neighbourhood) is performed by random walks, and global exploration depends on the randomization via Lévy flights. So, the step lengths are distributed according to the values generated in the real space via random walks and Lévy flights.

6.3 Travelling Salesman Problem

The so-called travelling salesman problem (TSP) forms a class of non-deterministic polynomial-time (NP) hard problems. The objective is to visit n cities or vertices once and exactly once so that the total distance travelled can be minimized. In essence, a TSP solution can be modelled as a weighted graph where the vertices of the graph correspond to cities and the graph edges correspond to connections or distances between cities. The weight of an edge is the corresponding connection distance between two cities. Thus, a tour or a solution now becomes a hamiltonian cycle and an optimal tour is the shortest hamiltonian cycle [21].

In our cuckoo search, an egg is the equivalent to a hamiltonian cycle and a nest is shown as an individual in the population with its own hamiltonian cycle. The solution is so shown as an integer (visiting order of the city) string. The coordinates of cities are fixed; however, the visiting order between the cities can be changed and thus forms a particular tour. The moves or perturbations used to change the order of visited cities are 2-opt moves [7] and double-bridge moves [25]. The 2-opt move method is used for small perturbations in the neighbourhood of the current solution, while large perturbations are made by double-bridge moves [27].

The computational results obtained by our discrete cuckoo search for 41 TSP benchmark instances based on the well-know TSPLIB library are summarized in Table 2 where the best results are highlighted in bold font.

The results of the discrete cuckoo search (DCS) have been compared with the results obtained by discrete particle swarm optimization (DPSO) [39] and the comparison is shown in Table 3. As we can see that the improved DCS obtained better results.

6.4 Scheduling Problem

Scheduling problems are another class of important combinational optimization problems. Briefly speaking, a schedule is an allocation of the operations to time

Table 2 Computational results of DCS algorithm for 41 TSP benchmark instances for TSPLIB

Instance	Opt	Best	Worst	Average	SD	PDav (%)	PDbest (%)	$C_{1\%}/C_{opt}$	temps
eil51	426	**426**	**426**	**426**	**0.00**	**0.00**	**0.00**	**30/30**	1.16
berlin52	7542	**7542**	**7542**	**7542**	**0.00**	**0.00**	**0.00**	**30/30**	0.09
st70	675	**675**	**675**	**675**	**0.00**	**0.00**	**0.00**	**30/30**	1.56
pr76	108159	**108159**	**108159**	**108159**	**0.00**	**0.00**	**0.00**	**30/30**	4.73
eil76	538	**538**	539	538.03	0.17	0.00	0.00	30/29	6.54
kroA100	21282	**21282**	**21282**	**21282**	0.00	**0.00**	**0.00**	**30/30**	2.70
kroB100	22141	**22141**	22157	22141.53	2.87	**0.00**	**0.00**	30/29	8.74
kroC100	20749	**20749**	**20749**	**20749**	**0.00**	**0.00**	**0.00**	**30/30**	3.36
kroD100	21294	**21294**	21389	21304.33	21.79	0.04	**0.00**	30/19	8.35
kroE100	22068	**22068**	22121	2281.26	18.50	0.06	**0.00**	30/18	14.18
eil101	629	**629**	633	630.43	1.14	0.22	**0.00**	30/6	18.74
lin105	14379	**14379**	**14379**	**14379**	**0.00**	**0.00**	**0.00**	**30/30**	5.01
pr107	44303	**44303**	44358	44307.06	12.90	0.00	**0.00**	30/27	12.89
pr124	59030	**59030**	**59030**	**59030**	**0.00**	**0.00**	**0.00**	**30/30**	3.36
bier127	118282	**118282**	118730	118359.63	12.73	0.06	**0.00**	30/18	25.50
ch130	6110	**6110**	6174	6135.96	21.24	0.42	**0.00**	28/7	23.12
pr136	96772	96790	97318	97009.26	134.43	0.24	0.01	30/0	35.82
pr144	58537	**58537**	**58537**	**58537**	**0.00**	**0.00**	**0.00**	**30/30**	2.96
ch150	6528	**6528**	6611	6549.9	20.51	0.33	**0.00**	29/10	27.74
kroA150	26524	**26524**	26767	26569.26	56.26	0.17	**0.00**	30/7	31.23
kroB150	26130	**26130**	26229	26159.3	34.72	0.11	**0.00**	30/5	33.01
pr152	73682	**73682**	**73682**	**73682**	**0.00**	**0.00**	**0.00**	**30/30**	14.86
rat195	2323	2324	2357	2341.86	8.49	0.81	0.04	20/0	57.25

(continued)

Table 2 (continued)

Instance	Opt	Best	Worst	Average	SD	PDav (%)	PDbest (%)	$C_{1\%}/C_{opt}$	temps
d198	15780	15781	15852	15807.66	17.02	0.17	0.00	30/0	59.95
kroA200	29368	29382	29886	29446.66	95.68	0.26	0.04	29/0	62.08
kroB200	29437	29448	29819	29542.49	92.17	0.29	0.03	28/0	64.06
ts225	126643	**126643**	126810	126659.23	44.59	0.01	**0.00**	30/26	47.51
tsp225	3916	**3916**	3997	3958.76	20.73	1.09	**0.00**	9/1	76.16
pr226	80369	**80369**	80620	80386.66	60.31	0.02	**0.00**	30/19	50.00
gil262	2378	2382	2418	2394.5	9.56	0.68	0.16	22/0	102.39
pr264	49135	**49135**	49692	49257.5	159.98	0.24	**0.00**	28/13	82.93
a280	2579	**2579**	2623	2592.33	11.86	0.51	**0.00**	25/4	115.57
pr299	48191	48207	48753	48470.53	131.79	0.58	0.03	27/0	138.20
lin318	42029	42125	42890	42434.73	185.43	0.96	0.22	15/0	156.17
rd400	15281	15447	15704	15533.73	60.56	1.65	1.08	0/0	264.94
fl417	11861	11873	11975	11910.53	20.45	0.41	0.10	30/0	274.59
pr439	107217	107447	109013	107960.5	438.15	0.69	0.21	22/0	308.75
rat575	6773	6896	7039	6956.73	35.74	2.71	1.81	0/0	506.67
rat783	8806	9043	9171	9109.26	38.09	3.44	2.69	0/0	968.66
pr1002	259045	266508	271660	268630.03	1126.86	3.70	2.88	0/0	1662.61
nrw1379	56638	58951	59837	59349.53	213.89	4.78	4.08	0/0	3160.47

Table 3 Comparison of experimental results of the DCS with DPSO [39] (TSP instances)

Instance	Opt	DPSO			Improved DCS		
		Best	Worst	PDAv (%)	Best	Worst	PDAv (%)
eil51	426	427	452	2.57	**426**	**426**	**0.00**
berlin52	7542	**7542**	8362	3.84	**7542**	**7542.0**	**0.00**
st70	675	**675**	742	3.34	**675**	**675**	**0.00**
pr76	108159	108280	124365	3.81	**108159**	**108159**	**0.00**
eil76	538	546	579	4.16	**538**	539	0.00

intervals on all machines that process the tasks or jobs. The main aim of a scheduling problem is to find a feasible schedule that minimizes the overall time span required to complete all jobs. There are many ways to generate a new solution, though a solution is typically a permutation of operations and operators used to move in the search space by swapping, insertion and inverse moves. Such moves can be associated with random sequences following a Lévy distribution [30]. A local search step in the neighbourhood is achieved by applying one of these moves and an explorative step is represented by a series or combination of these three move operators.

In the cuckoo search, one of the approaches is to encode a nest or sequence of jobs (solution) as a collection of sets of two-digit numbers (series of integers). To move from a current nest to a new one, Lévy flight is performed to generate and add big integer number to the given solution [6].

An important class of scheduling problems is the so-called job shop scheduling problem (JSSP). We have solved some JSSP benchmark instances using the presented discrete cuckoo search and compared them with the results obtained by PSO, and the comparison is summarized in Table 4 where '–' means that the results for PSO were not available from the literature. As we can see from the table, DCS outperforms PSO for all the instances and in many cases (highlighted in bold font) the results obtained by DCS are also global optima.

6.5 Vehicle Routing Problem

The vehicle routing problem (VRP) is another class of important combinational optimization. The main aim of a VRP is to find the minimal set of routes for a fleet of vehicles so that all customer demands are satisfied, subject to the constraints that each customer is visited by exactly one vehicle and that the capacity of each vehicle is not exceeded [42]. One way to represent a solution is to use an integer string. For example, in the cuckoo search, each nest in the string is the integer node number assigned to that customer originally. The sequence of the particle in the string is the order of visiting these customers. Operators performed to move in the solution space are node swapping and insertion operators [50].

Table 4 Comparison of PSO [22] and DCS algorithms for JSSP instances

	DCS			PSO		
Instance (opt)	Best	Average	PDAv (%)	Best	Average	PDAv (%)
Abz5 (1234)	1239	1239.6	0.00	–	–	–
Abz6 (943)	943	946.4	0.00	–	–	–
Ft06 (55)	**55**	**55**	**0.00**	55	56.1	0.02
Ft10 (930)	945	966.8	0.03	985	1035.6	0.11
Ft20 (1165)	1173	1178.8	0.01	1208	1266.9	0.08
La01 (666)	**666**	**666**	**0.00**	666	668.6	0.00
La02 (655)	**655**	**655**	**0.00**	–	–	–
La03 (597)	604	607	0.01	–	–	–
La04 (590)	**590**	**590**	**0.00**	–	–	–
La06 (926)	**926**	**926**	**0.00**	926	926	0.00
La11 (1222)	**1222**	**1222**	**0.00**	1222	1222	0.00
La16 (945)	946	967.7	0.02	956	986.9	0.04
La21 (1046)	1055	1070.1	0.02	1102	1128.4	0.07
La26 (1218)	**1218**	**1218**	**0.00**	1263	1312.6	0.07
La31 (1784)	**1784**	**1784**	**0.00**	1789	1830.4	0.02
La36 (1268)	1297	1313.6	0.03	1373	1406.2	0.10

Obviously, there are other combinatorial optimization problems. As the literature of metaheuristics continues to expand, there is no doubt that more applications will appear in the literature. The review of such studies will be the topic of future work.

7 Conclusion

In this chapter, we have focused on the understanding on how to carry out search in the combinatorial search space from the inspiration from nature. The essence of a nature-inspired algorithm is to generate good solutions (ideally always better solutions) based on the current solutions and the moves performed by the algorithm, and in the end, after a fixed number of iterations (ideally a very small number of iterations), the global optimal solution can be found, though this is rarely possible for most combinatorial optimization problems. In fact, there are no efficient methods for solving NP-hard problems, though nature-inspired metaheuristic algorithms form a promising set of alternative techniques that can obtain good solutions in many cases in a practically acceptable time scale.

Obviously, despite the active research in the area concerning both combinatorial optimization and metaheuristic methods, there are still many things to be done and there are many key open problems to be resolved. For example, in general, there still lacks a framework for mathematical analysis of these algorithms and it is not clear if these algorithms can be used to solve truly large-scale problems. In addition, each algorithm has some algorithm-dependent parameters and the values of these parameters can affect the performance of an algorithm. Thus, how to tune these parameters quickly so that the algorithm can perform to the best possible degree is still an open problem. Furthermore, detailed sensitivity studies to test the robustness of an algorithm is also useful, so is the theoretical analysis of convergence of metaheuristic algorithms. All these will inspire more research in this area of active research and it can be expected that more research will appear in the future.

References

1. Arora, S., Barak, B.: Computational Complexity: A Modern Approach. Cambridge University Press (2009)
2. Bak, P.: How Nature Works: The Science of Self-organized Criticality. Springer Science & Business Media (2013)
3. Blum, C., Roli, A.: Metaheuristics in combinatorial optimization: overview and conceptual comparison. ACM Comput. Surv. (CSUR) **35**(3), 268–308 (2003)
4. Briers, R.: Ecology: from individuals to ecosystems. Freshwater Biol. **51**(9), 1787–1788 (2006)
5. Brown, C.T., Liebovitch, L.S., Glendon, R.: Lévy flights in dobe ju/hoansi foraging patterns. Human Ecol. **35**(1), 129–138 (2007)
6. Burnwal, S., Deb, S.: Scheduling optimization of flexible manufacturing system using cuckoo search-based approach. Int. J. Adv. Manuf. Technol. **64**(5–8), 951–959 (2013)
7. Croes, G.A.: A method for solving traveling-salesman problems. Oper. Res. **6**(6), 791–812 (1958)
8. Davies, N., Brooke, M.D.L.: An experimental study of co-evolution between the cuckoo, cuculus canorus, and its hosts. ii. host egg markings, chick discrimination and general discussion. J. Anim. Ecol. 225–236 (1989)
9. Dejam, S., Sadeghzadeh, M., Mirabedini, S.J.: Combining cuckoo and tabu algorithms for solving quadratic assignment problems. J. Acad. Appl. Stud. **2**(12), 1–8 (2012)
10. Garey, M.R., Johnson, D.S.: Computers and Intractability: A Guide to np-completeness (1979)
11. Garnier, S., Gautrais, J., Theraulaz, G.: The biological principles of swarm intelligence. Swarm Intell. **1**(1), 3–31 (2007)
12. Gherboudj, A., Layeb, A., Chikhi, S.: Solving 0–1 knapsack problems by a discrete binary version of cuckoo search algorithm. Int. J. Bio-Inspir. Comput. **4**(4), 229–236 (2012)
13. Glover, F., Kochenberger, G.A.: Handbook of Metaheuristics. Springer Science & Business Media (2003)
14. Glover, F., Laguna, M: Tabu Search. Springer (2013)
15. Hall, B. et al.: Strickberger's Evolution. Jones & Bartlett Learning (2008)
16. Jones, C.G., Lawton, J.H., Shachak, M.: Organisms as ecosystem engineers. In Ecosystem Management, pp. 130–147. Springer (1996)
17. Kanagaraj, G., Ponnambalam, S., Jawahar, N.: A hybrid cuckoo search and genetic algorithm for reliability-redundancy allocation problems. Comput. Ind. Eng. **66**(4), 1115–1124 (2013)
18. Kennedy, J., Kennedy, J.F., Eberhart, R.C., Shi, Y.: Swarm Intelligence. Morgan Kaufmann (2001)
19. Kirkpatrick, S., Gelatt, C.D., Vecchi, M.P., et al.: Optimization by simulated annealing. Science **220**(4598), 671–680 (1983)
20. Lawler, E.L.: The quadratic assignment problem. Manage. Sci. **9**(4), 586–599 (1963)
21. Lawler, E.L.: The traveling salesman problem: a guided tour of combinatorial optimization. Wiley-interscience series in discrete mathematics (1985)

22. Lin, T.-L., Horng, S.-J., Kao, T.-W., Chen, Y.-H., Run, R.-S., Chen, R.-J., Lai, J.-L., Kuo, I.-H.: An efficient job-shop scheduling algorithm based on particle swarm optimization. Expert Syst. Appl. **37**(3), 2629–2636 (2010)
23. Lozano, M., García-Martínez, C.: Hybrid metaheuristics with evolutionary algorithms specializing in intensification and diversification: overview and progress report. Comput. Oper. Res. **37**(3), 481–497 (2010)
24. Martello, S., Toth, P.: Knapsack Problems: Algorithms and Computer Implementations. Wiley (1990)
25. Martin, O., Otto, S.W., Felten, E.W.: Large-step markov chains for the traveling salesman problem. Complex Syst. **5**(3), 299–326 (1991)
26. Neumann, F., Witt, C: Minimum Spanning Trees. Springer (2010)
27. Ouaarab, A., Ahiod, B., Yang, X.-S.: Discrete cuckoo search algorithm for the travelling salesman problem. Neural Comput. Appl. **24**(7–8), 1659–1669 (2014)
28. Ouaarab, A., Ahiod, B., Yang, X.-S.: Improved and discrete cuckoo search for solving the travelling salesman problem. In: Cuckoo Search and Firefly Algorithm, pp. 63–84. Springer (2014)
29. Ouaarab, A., Ahiod, B., Yang, X.-S.: Discrete cuckoo search applied to job shop scheduling proble. In: Recent Advances in Swarm Intelligence and Evolutionary Computation, pp. 121–137. Springer (2015)
30. Ouaarab, A., Ahiod, B., Yang, X.-S., Abbad, M.: Discrete cuckoo search algorithm for job shop scheduling problem. In: IEEE International Symposium on Intelligent Control (ISIC), 2014, pp. 1872–1876. IEEE (2014)
31. Ouaarab, A., Ahiod, B., Yang, X.-S., Abbad, M.: Discrete cuckoo search for the quadratic assignment problem. In: The XI Metaheuristics International Conference (2015)
32. Papadimitriou, C.H., Steiglitz, K.: Combinatorial Optimization: Algorithms and Complexity. Courier Corporation (1998)
33. Parker, R.G., Rardin, R.L.: Discrete Optimization. Elsevier (2014)
34. Payne, R.B.: The ecology of brood parasitism in birds. Ann. Rev. Ecol. Syst. 1–28 (1977)
35. Payne, R.B., Sorensen, M.D.: The Cuckoos, vol. 15. Oxford University Press (2005)
36. Pinedo, M.L.: Scheduling: Theory, Algorithms, and Systems. Springer Science & Business Media (2012)
37. Rochat, Y., Taillard, É.D.: Probabilistic diversification and intensification in local search for vehicle routing. J. Heuristics **1**(1), 147–167 (1995)
38. Schrijver, A.: Combinatorial Optimization: Polyhedra and Efficiency, vol. 24. Springer Science & Business Media (2003)
39. Shi, X., Liang, Y., Lee, H., Lu, C., Wang, Q.: Particle swarm optimization-based algorithms for tsp and generalized tsp. Inform. Process. Lett. **103**(5), 169–176 (2007)
40. Shlesinger, M.F., Zaslavsky, G.M., Frisch, U.: Lévy flights and related topics in physics. In: Levy Flights and Related Topics in Physics, vol. 450 (1995)
41. Talbi, E.-G.: Metaheuristics: From Design to Implementation, vol. 74. Wiley (2009)
42. Toth, P., Vigo, D.: The vehicle routing problem. Soc. Ind. Appl. Math. (2001)
43. Winfree, R.: Cuckoos, cowbirds and the persistence of brood parasitism. Trends Ecol. Evol. **14**(9), 338–343 (1999)
44. Wolsey, L.A., Nemhauser, G.L.: Integer and Combinatorial Optimization. Wiley (2014)
45. Yang, X.-S.: Firefly algorithm, levy flights and global optimization. In: Research and Development in Intelligent Systems XXVI, pp. 209–218. Springer (2010)
46. Yang, X.-S.: Swarm-Based Metaheuristic Algorithms and No-free-Lunch Theorems. INTECH Open Access Publisher (2012)
47. Yang, X.-S.: Nature-Inspired Optimization Algorithms. Elsevier (2014)
48. Yang, X.-S., Deb, S.: Cuckoo search via lévy flights. In: World Congress on Nature & Biologically Inspired Computing, 2009. NaBIC 2009. pp. 210–214. IEEE (2009)
49. Yang, X.-S., Deb, S.: Engineering optimisation by cuckoo search. Int. J. Math. Model. Numer. Optim. **1**(4), 330–343 (2010)
50. Zheng, H., Zhou, Y., Luo, Q.: A hybrid cuckoo search algorithm-grasp for vehicle routing problem. J. Convergence Inform. Technol. **8**(3) (2013)

Clustering Optimization for WSN Based on Nature-Inspired Algorithms

Marwa Sharawi and Eid Emary

Abstract This chapter presents a set of newly proposed swarm intelligent methods and applies these method in the domain of wireless sensor network (WSN) for the purpose of cluster head selection. Life time of WSNs is always the main performance goal. Cluster head (CH) selection is one of the factors affecting the life time of WSNs and hence it is a very promising area of research. Swarm-intelligence is a very hot area of research which mimics natural behavior to solve optimization problems. This chapter formulates the CH selection problem as an optimization problem and tackles this problem using different emergent swarm optimizers. The proposed formulation is assessed using different performance indicators and is compared against one of the very common CH selection methods namely LEACH.

1 Wireless Sensor Network

Wireless Sensor Network (WSN) is a one type of an ad-hoc distributed networking system. It consists of spatially distributed self-dependent large number of embedded numerous physical tiny nodes called sensors. It is scientifically defined as a collective system of distributed, homogenous, autonomous and self-organized tiny sensors' nodes [1]. They are in charge of some different activities, responsible for sensing, processing and communicating in WSNs [2]. WSN has an innovative paradigm that enables it to be widely used in many areas. It used to efficiently monitor physical or

M. Sharawi (✉)
Faculty of Computer Studies, Arab Open University, Cairo, Egypt
e-mail: m.sharawi@aou.edu.eg

E. Emary
Faculty of Computers and Information, Cairo University, Giza, Egypt
e-mail: eidemary@yahoo.com

© Springer International Publishing Switzerland 2016
X.-S. Yang (ed.), *Nature-Inspired Computation in Engineering*,
Studies in Computational Intelligence 637, DOI 10.1007/978-3-319-30235-5_6

environmental conditions, such as temperature, sound, vibration, pressure, motion at different locations. The sensor nodes are the most important element in WSNs. They are battery operated devices; they derives all their computations through small embedded electronic circuits. Beside these circuits there is radio transceiver along with an antenna and a micro-controller [2].

1.1   Optimization Challenges in Wireless Sensor Network

Challenges of wireless sensor networks under-optimization in the field of research have been of global concern. Lifetime extension is considered to be the universal and most dominant challenge for this type of networks. Routing protocols based on energy optimization were and are still addressed as conventional solution to optimize WSNs. However, because of the scarcity in sensors' power and the ambiguity and uncertainty of the data in complex WSN's environments nowadays, classical routing protocols do not capable anymore to cope efficiently [3]. The classical existing WSN routing protocols suffer a crucial problem. Once there is an optimal route investigated and determined by the chosen routing protocol, the network keeps using it for every transmission [4]. This gradually drives to massive loss in the route nodes' energy. Consequently causes network partitioning because of the dramatic loss of the nodes power. Clustering approaches and hierarchical routing protocols have been proposed as a solution to optimize the WSN's lifetime. It transforms the WSN's topological architecture from homogenous to heterogeneous deployment. Sensor nodes are organized into different clusters, with a cluster head in each cluster executing the data collecting and transmitting tasks for other member nodes. This optimizes the power utilization of the WSN overall to extend the network lifetime [2]. Low-energy Adaptive Clustering Hierarchy (LEACH) is a classical version of hierarchical routing protocol.

1.2   Low-Energy Adaptive Clustering Hierarchy Algorithm

LEACH divides the communication process into rounds with each round including a set-up phase and a steady-state phase. It relies on cluster formation and selection of a head for each cluster. WSN's nodes are grouped into a number of clusters randomly. Based on the available energy in each sensor node; nodes elect themselves to be chosen as a cluster head (CH) [5]. LEACH acts as follows; the randomly scattered nodes in homogenous WSNs are grouped into clusters. Election process takes place to select cluster head based on the probability factor of the available energy in each node. The latter elected cluster head is responsible for the data aggregation from cluster nodes to base station [6]. The communication occurs between all the cluster members and cluster heads. Similarly cluster heads communicate to the sink/base station (BS) node. Although LEACH protocol prolongs the network lifetime in contrast to plane multi-hop routing and static routing, it still suffers the following problems [7]:

- The cluster heads are elected randomly so it is possible that two or even more cluster heads are set to be very close to each other. This neither guarantees the optimal number nor the good distribution of cluster heads.
- When multiple cluster heads are randomly selected within a small area, a big extra energy loss occurs as the amount of lost energy is approximately proportional to the number of cluster heads in the area.
- The nodes with low remaining energy have the same priority to be a cluster head as the node with high remaining energy as long they are not elected as a cluster head in previous rounds. Therefore, those nodes with less remaining energy may be chosen as the cluster heads which will result that these nodes may die first.
- The cluster heads communicate with the base station in single-hop mode. Thus, prevents LEACH cannot be used in large-scale wireless sensor networks for the limit effective communication range of the sensor nodes.
- LEACH assumes a homogeneous distribution of sensor nodes in the given area. This scenario is not very realistic because at some cases of the random deployments most of the sensor nodes are naturally grouped together around one or two cluster head nodes. Therefore, one cluster-head may have more nodes close to it than the other cluster heads. LEACHs cluster formation algorithm will end up by assigning more cluster member nodes to this cluster head node. This could make these cluster head nodes run out of energy quickly.
- LEACH restricts and dominates the selection of the next cluster head to only nodes that weren't selected as cluster heads in the previous $1/P$ rounds. This limits the chances of higher energy node to be selected as a cluster head.
- LEACH selects the P percent of nodes to act as a cluster heads in the cluster formation process. However, as P depends on the number of the available sensor nodes in the network, the death of sensor nodes affect the value of P. Hence, the number of nominated cluster head nodes will decrease.
- When P become less than 1, all the available nodes in the network are forced to communicate directly with the BS.

2 Proposed Clustering Optimization Model

In this section we formulate hierarchical clustering and routing model for WSN as a single objective optimization problem. The following subsections describe the network model, energy model, cluster formation, cluster head selection and the data transmission. All with its own applied parameters.

2.1   Network Model

The proposed model is based on a homogenous squared range WSNs' deployments lies on the following assumptions:

- Number of nodes N, are randomly distributed across the (M*N) region and all are energy constrained.
- All nodes are stationary and constrained that no longer movement for any node after deployment (*static deployment*).
- The *N* stationary sensor nodes are scattered and distributed uniformly in the deployed region.
- All the deployed nodes have the same initial energy level (*uniform initial energy*).
- Sensors are regularly sending data to destination node.
- Nodes are located near to each other, and have correlated data.
- Numbers of neighbored nodes are grouped into cluster.
- The base station (*BS*) is a high-energy node, fixed, motionless and stationary central device that is located far outside the WSN's deployment sensing region in the point (X_{sink}, Y_{sink}).

In each cluster; nodes usually sense and collect data from the surrounding environment, they always have data to send. They have a fixed sensing rate and directly connected to the corresponding cluster head. Sensor nodes transmit their data packets directly to their cluster head. Cluster heads are responsible for data aggregation. Each cluster head collects data from all of its cluster nodes and performs some necessary iteration for compression. It is connected directly to the sink node *base station* to forward the aggregated data packets. Nodes die only when they are out of energy.

2.2   Energy Model

For calculating the WSN consumed energy, we adopt the first order radio frequency energy consumption model [5]. This model can be divided into two parts according to the distance between the sender and the receiver, the first part is the free-space model while the other is the multi-path fading model. The following energy assumptions are applied in the proposed model:

- Initialize all the nodes with their properties and set their initial energies based on the first order radio energy model.
- Nodes are transmitting message with k bits through a distance d and hence consume energy.
- Assumes that we apply symmetrical communication channel.
- Energy consumed for k bits length message sent between two nodes operated with a distance d can be calculated based on the first order radio energy model [5].

2.3 Cluster Formation

Cluster formation process is based on the intra-cluster euclidean distances between nodes based on the Eq. 1. So the objective is to find the number of cluster centers (N) that minimizes this equation.

$$J = \sum_{j=1}^{x} \sum_{i=1}^{k} \|x_i^j - c_j\|^2 \tag{1}$$

where x_i^j is the senor node i that belongs to the cluster j. c_j is the cluster head node for the cluster j. $(x_i^j - c_j)$ is the distance between sensor node position and cluster head.

2.4 Cluster Head Selection

For every cluster obtained from the proposed clustering model a candidate cluster head (CH) is nominated/selected. The CH_i is selected as the node inside the cluster i with the most remaining energy; this can be calculated as follows:

$$CH_i = max_{nodes \in C_i} Remaining\ Energy \tag{2}$$

where CH_i is selected as the node inside the cluster i with the most remaining energy.

2.5 Data Transmission

This subsection describes the data transmission process after clusters formation. It is responsible to transfer data packets from all the entire nodes in each cluster to the base station through each cluster head. This phase stands for the following assumptions:

- Each cluster head is directly connected to the sink node (*base station*) which is known as a station node with a very high energy and computation capabilities. It is positioned far away from the field of WSN deployment in the point (X_{sink}, Y_{sink}).
- Data packets are sent directly from nodes to CH.
- Each CH receives data from all of its cluster nodes and performs some necessary iteration for compression. It is directly connected to the sink node (*base station*) to forward the aggregated data packets.
- Nodes die only when they are out of energy.

All the steps that summarize the proposed clustering optimization model are shown in Algorithm 1.

Input: WSN Architecture,Sensor Nodes' Architecture,Round Architecture.
Output: WSN's Cluster

1. *WSN Architecture* Create a homogenous WSN squared architecture ($M * N$), locate the base station and set the WSN's energy model (Initial Energy, Energy for transferring of each bit, Transmit Amplifier energy and Data Aggregation Energy)
2. *Sensor Nodes' Architecture* Create the sensor node model randomly based on uniform distribution of the energy constrained stationary N sensors' devices in the WSN deployed field. This is a static deployment.
3. *Round Architecture* Create the round architecture for specific round parameters such as the maximum allowable number of rounds, the size of data packets sent from cluster head to base station and the size of data packets sent from nodes to CH.
4. Find the best solution.
5. $t=0$.
6. while ($t \leq T$)
   foreach *round$_i$* **do**

   - *Cluster Formation* based on equation 1.
   - *Cluster Head Selection* based on equation 2.

   end
7. *Data Transmission*
8. Nodes die only when they are out of energy. end while

Algorithm 1: Proposed Clustering Optimization Model

3   Natural Inspired Optimization Algorithms

This section presents the applied nature inspired algorithm applied to solve the clustering problem in WSN.

3.1   Bat Algorithm

Bat Algorithm (BA) is a population-based optimization method. It solves nonlinear global optimization problems with single or multi-objective functions. It is exposed as a meta-heuristic swarm intelligence optimization method developed for the global numerical optimization. This algorithm is naturally inspired from the social behavior of bats [8]. The capability of echo-locations of these bats composed a great competent manner to detect prey, avoid obstacles and to locate their roosting crevices in the dark based on sensing distances.

This algorithm starts with the initialization of all the echolocation system variables. Initial location of all bat swarm should be initialized as initial solutions. Values of *pulse* and *loudness* are set randomly as well as the value of the *frequency*. Within a number of iteration bats try to find the best optimized solution(s) moves from their initial state solutions toward the optimal global best solution(s). Bats solutions are automatically updated in the sense of finding better solution. Both pulse emission rate and loudness are updating gradually as closer as bats reach their best solution(s). Solutions keep updated as a result of the continuous flying iterations till the termination criteria are satisfied. Finally, when all criteria successfully met the best so far solution is visualized.

Bats adjust their position according to the Eqs. 3, 4, and 23.

$$f_i = f_{min} + (f_{max} - f_{min})\beta \tag{3}$$

$$v_i^t = v_i^{t-1} + (x_i^{t-1} - x*)f_i \tag{4}$$

$$x_i^t = x_i^{t-1} + v_i^t \tag{5}$$

where β is a random vector in the range [0, 1] drawn from uniform distribution. $x*$ is the current global best location. f_{min} and f_{max} represent the minimum and maximum frequency defined depending on the problem. v_i represents the velocity vector. Probabilistically a local search is to be performed using a random walk as in the Eq. 22.

$$x_{new} = x_{old} + \varepsilon A^t \tag{6}$$

where A^t is the average loudness of all bats at this time and ε is a random number uniformly drawn from [−1 1]. The updating of the loudness is performed using the Eq. 7.

$$A_i^{t+1} = \alpha A^t \tag{7}$$

where α is a constant selected experimentally. r_i controls the application of the local search as shown on the algorithm and is updated using the Eq. 8.

$$r_i^{t+1} = r_i^0[1 - exp(-\gamma t)] \tag{8}$$

where r_i^0 is the initial pulse emission rate and γ is a constant greater than 0.

3.2 Flower Pollination Algorithm

Flower Pollination Algorithm (FPA) is a recently invented optimization algorithm. It is inherited from the natural inspiration of pollination process. Flower pollination is typically associated with the transfer of pollen, and such transfer is often linked with pollinators such as insects, birds, bats and other animals [9]. It mimics the process of flowering planets reproduction via pollination. As pollinators are mainly responsible for transferring pollens among flowers, pollination

may occur in either local or global flow [10]. Pollination can be achieved by
Self-pollination or *cross-pollination*; local and global search in the artificial algorithm. *Cross-pollination*, or allogamy, means pollination can occur from pollen of
a flower of different plant. Biotic, cross-pollination may occur at long distance, and
the pollinators such as bees, bats, birds and flies can fly a long distance, thus they
can considered as the global pollination; global search in the artificial algorithm. In
addition, bees and birds may behave as Levy flight behavior [11], with jump or fly
distance steps obey a Levy distribution. Furthermore, flower constancy can be used
an increment step using the similarity or difference of two flowers [9].

The simple flower pollination model assumes that each plant has only one flower,
and each flower only produces one pollen gamete. Thus, there is no need to distinguish
a pollen gamete, a flower, a plant or solution to a problem [10]. According to the
rules above, the flower pollination optimization algorithm (FA) can be represented
mathematically as follows:

In the *global pollination*, flower pollens are carried by pollinators such as insects,
and pollens can travel over a long distance. This ensures the pollination and reproduction of the most fittest, and thus we represent the most fittest as g_*. The first rule
can be formulated as bee is updated using the Eq. 9 [17, 18]:

$$X_i^{t+1} = X_i^t + L(X_i^t - g_*) \tag{9}$$

where X_i^t is the solution vector i at iteration t and g_* is the current best solution. The
parameter L is the strength of the pollination which is the step size randomly drawn
from Levy distribution [19].

For *Self-pollination* is the fertilization of one flower, such as peach flowers, from
pollen of the same flower or different flowers of the same plant, which often occurs
when there is no reliable pollinator available. The local pollination and flower constancy can be represented as in Eq. 10 [17, 18]

$$X_i^{t+1} = X_i^t + \varepsilon(X_j^t - g_k^t) \tag{10}$$

where X_j^t and X_k^t are solution vectors drawn randomly from the solution set. The
parameter ε is drawn from uniform distribution in the range from 0 to 1.

Due to the physical proximity and other factors such as wind, local pollination
can have a significant fraction p in the overall pollination activities.

3.3 Antlion Algorithm (AL)

Antlion Algorithm (AL) is a recently proposed optimization algorithm by Mirjalili
[12]. It mimics the hunting mechanism of antlions in nature. Antlions (doodle-
bugs) belong to the Myrmeleontidae family and Neuroptera order [12]. They mostly
hunt in larvae and the adulthood period is for reproduction. An antlion larvae digs

a cone-shaped hole in sand by moving along a circular path and throwing out sands with by its huge jaw. After digging the trap, the larvae hides underneath the bottom of the cone and waits for insects (preferably ant) to be trapped in the hole. Once the antlion realizes that a prey is in the trap, then antlion tries to catch its prey. However, insects usually are not caught immediately and try to escape from the trap. In this case, antlions intelligently throw sands towards to edge of the hole to slide the prey into the bottom of the hole. When a prey is caught into the jaw, it is pulled under the soil and consumed. After consuming the prey, antlions throw the leftovers outside the hole and amend the hole for the next hunt.

Based on the above description of antlions Mirjalili set the following condition during optimization:

- Preys; ants, move around the search space using different random walks.
- Random walks are affected by the traps of antlions.
- Antlions can build holes proportional to their fitness (the higher fitness, the larger hole).
- Antlions with larger holes have the higher probability to catch ants.
- Each ant can be caught by an antlion in each iteration and the elite (fittest antlion).
- The range of random walk is decreased adaptively to simulate sliding ants towards antlions.
- If an ant becomes fitter than an antlion, this means that it is caught and pulled under the sand by the antlion.
- An antlion repositions itself to the latest caught prey and builds a hole to improve its change of catching another prey after each hunt.

The antlion optimizer applies the following steps to individual antlion:

1. **Building trap**: Roulette wheel is used to model the hunting capability of antlions. Ants are assumed to be trapped in only one *selected* antlion. The ALO algorithm is required to utilize a roulette wheel operator for selecting antlions based of their fitness during optimization.

2. **Catching prey and re-building the hole**: This is the final stage in hunting where the antlion consumes the ant. For mimicking this process, it is assumed that catching prey occur when ants becomes fitter (goes inside sand) than its corresponding antlion. An antlion is then required to update its position to the latest position of the hunted ant to enhance its chance of catching new prey. The following Eq. 11 is proposed in this regard:

$$Antlion_j^t = Ant_i^t \text{ If } f(Ant_i^t) \text{ is better than } f(antlion_j^t), \qquad (11)$$

where t shows the current iteration, $Antlion_j^t$ shows the position of selected j-th antlion at t-th iteration, and Ant_i^t indicates the position of i-th ant at t-th iteration.

According to the algorithm, the antlion optimizer applies the following steps to individual ant:

1. **Sliding ants towards antlion**: Antlions shoot sands outwards the center of the hole once they realize that an ant is in the trap. This behaviour slides down the trapped ant that is trying to escape. For mathematically modelling this behaviour, the radius of antss random walks hyper-sphere is *decreased* adaptively; see the Eqs. 12–14.

$$c^t = \frac{c^t}{I},$$ (12)

where c^t is the minimum of all variables at t'th iteration and I is a ratio.

$$d^t = \frac{d^t}{I},$$ (13)

where d^t is the maximum of all variables at t'th iteration and I is a ratio that is defined as:

$$I = 10^w \frac{t}{T},$$ (14)

where t is the current iteration, T is the maximum number of iterations, and w is a constant defined based on the current iteration ($w = 2$ when $t > 0.1T$, $w = 3$ when $t > 0.5T$, $w = 4$ when $t > 0.75T$, $w = 5$ when $t > 0.9T$, and $w = 6$ when $t > 0.95T$). Basically, the constant w can adjust the accuracy level of exploitation.

2. **Trapping in antlion's holes**: By modeling the sliding of prey towards the selected antlion, the slide ant is trapped in the selected antlion's hole. In another words the walk of the ant becomes bounded by the position of the selected antlion which can be modeled by changing the range of the antś random walk to the antlions position as in the Eqs. 15, and 16.

$$c_i^t = c^t + Antlion_j^t,$$ (15)

$$d_i^t = d^t + Antlion_j^t,$$ (16)

where c^t is the minimum of all variables at t'th iteration, d^t indicates the vector including the maximum of all variables at t'th iteration, c_i^t is the minimum of all variables for i-th ant, d_j^t is the maximum of all variables for i'th ant, and $Antlion_j^t$ shows the position of the selected j'th antlion at t'th iteration.

3. **Random walks of ants**: Random walks are all based on the Eq. 17.

$$X(t) = [0, cumsum(2r(t_1) - 1); cumsum(2r(t_2) - 1); ...; cumsum(2r(t_T) - 1)], \quad (17)$$

where $cumsum$ calculates the cumulative sum, T is the maximum number of iteration, t shows the step of random walk (iteration in this study), and $r(t)$ is a stochastic function defined in Eq. 18.

$$r(t) = \begin{cases} 1 \text{ if } rand > 0.50 \quad \text{if } rand \leq 0.5, \end{cases} \quad (18)$$

where t shows the step of random walk (iteration in this study) and rand is a random number generated with uniform distribution in the interval of [0, 1].

In order to keep the random walks inside the search space, they are normalized using the Eq. 19 (minmax normalization):

$$X_i^t = \frac{(X_i^t - a_i) \times (d_i - c_i^t)}{(b_i^t - a_i)} + c_i, \quad (19)$$

where a_i is the minimum of random walk of i'th variable, b_i is the maximum of random walk in i'th variable, c_i^t is the minimum of i'th variable at t'th iteration, and d_i^t indicates the maximum of i'th variable at t'th iteration.

4. **Elitism**: To maintain the best solution(s) across iterations elitism should be applied. In here, the random walk of ant is guided by the selected antlion as well as the elite antlion and hence the repositioning of a given ant takes the form of average of both the random walks; see Eq. 20.

$$Ant_i^t = \frac{R_A^t + R_E^t}{2}, \quad (20)$$

where R_A^t is the random walk around the roulette wheel selected antlion and R_E^t is the random walk around the elite antlion.

4   Experiments

This section illustrates the experiment and applied parameters of the proposed model. Simulation of the proposed model is examined with four different WSN deployments (A, B, C and D) using MATLAB runtime environment. The first order radio model in [5] is applied for energy computations in the proposed simulation. the parameters of the applied nature inspired algorithms and LEACH parameters are set optimally based on previous best practices.

4.1 Parameters Setting

Table 1 sets the parameters of these different WSNs' deployments. Network ID, Network area, number of nodes and base station location are mentioned [13]. Parameters for WSN deployment (B) are applied based on the standard previous best practices and researches for WSN's clustering protocols and cluster head selection mechanisms [14]. However, the other mentioned deployments (A, C and D) are applied to examine the scalability and validity of the proposed model for different deployments comparatively with the LEACH protocol [15]. In each cluster the length of data packets sent from these sensor nodes to cluster head is 200 bits in our model. While the Length of packet sent from each CH to BS is 6400 bits. They are communicating directly in one single hop with the base station.

Table 2 lists the applied parameters of the first order energy radio model. These are standard parameters commonly used in WSNs' environments based on the common technical and mechanical devices such as sensors and amplifiers [5].

Table 3 sets a maximum allowable number of rounds in LEACH as well as the percentage of nominated cluster heads [100] as well as the parameters of BA and FPA. Parameters are set randomly within a specific range based on parametric study to find the most appropriate parameters' values for each application. From our simulations,

Table 1 Network model parameters

ID	Area $(M * N)$ (m^2)	Number of nodes (N)	Base station location (X_{sink}, Y_{sink})
A	(70, 70)	25	35,145
B	(100, 100)	100	50,175
C	(250, 250)	150	125,325
D	(300, 300)	175	150,375

Table 2 Radio model parameters

Parameter	Value
Initial energy	0.5 J/node
Minimum energy	0.0001 J
Packet length from CH to BS	6400 bits
Packet length from sensor node to CH	200 bits
Electronics of transmitter	50 nJ/bit
Electronics of receiver	50 nJ/bit
Data aggregation energy	50 nJ/bit
Transmitter amplifier	10 pJ/bit/m^2
Transmitter amplifier	0.0013 pJ/bit/m^4
Network threshold	0.1 of nodes be alive

Table 3 Optimization parameters

Algorithms	Parameters	
LEACH	Maximum allowable number of rounds (r_i)	9999
	Percentage of nominated cluster heads (P)	0.5
BA	Population size (N)	No. nodes
	Number of generations	100
	Loudness (A_i)	0.5
	Pulse emission rate (r_i)	0.4
	Minimum frequency (f_{min})	0
	Maximum frequency (f_{max})	1
FPA	The standard gamma function (γ)	1.5
	The local random walk (ε)	[0, 1]
	The switch probability (p)	0.8

we found that the listed values of parameters work better for WSN's cluster formation and achieve good performance [10].

4.2 Performance Metrics

The following metrics are used to evaluate the proposed model.

4.2.1 WSN Lifetime

The WSN lifetime metric is an indication to the maximum coverage of the sensor network since its initial deployment to the first sensor node loss. The average Network life time of the Network is measured as the following [16]:

$$Average\ Network\ Lifetime = \frac{E_{Initial} - E_{Expected\ Wasted}}{E_{C\ Network} - E_{C\ Sensor}} \quad (21)$$

where $E_{Initial}$ is the initial energy. $E_{C\ Network}$ is the consumed energy by the network. $E_{C\ Sensor}$ is the consumed energy by the sensor. $E_{Expected\ Wasted}$ is the expected wasted energy.

Number of dead nodes will be used to measure the lifetime of WSN in the proposed model, this includes:

- **The death of first node**: this measure the time period since the WSN deployment of nodes to the death of the first node. This time interval measures the start of the WSN coverage reliability degradation.

- **The death of 50 % of the nodes**: this measure the time period since the WSN deployment of nodes to the 50 % node deaths. The metric indicates the consumed amount of the initial energy.
- **The death of all the nodes**: this measure the time period since the WSN deployment of nodes to the death of the last node. It measure the WSN lifetime.

4.2.2 Residual Energy

Residual energy is the available amount of energy in WSN's nodes. The residual energy per round in each cluster can be calculated as follow:

$$Residual\ Energy = E_{Total} - E_{Consumed} \tag{22}$$

where E_{Total} is the total energy per round in each cluster. $E_{Consumed}$ is the energy consumption per round in each cluster.

4.2.3 Network Throughput

The throughput reflects the effective network capacity. It is defined as the total number of bits successfully delivered at the destination in a given period of time. The ratio of total data received by a receiver from a sender within a time the last packet received by receiver measures in bit/s and byte/s [16]. Number of packets sent to BS that measures the performance of data gathering in WSN and ensures the success of data packets transmission across the network. It traces the number of received packets to the sink/Base Station node. Increasing the number of these packets over the rounds shows the efficiency of the proposed technique. Optimization of throughput is mainly the WSN data aggregation efficiency. It can be expressed mathematically as in [16]:

$$Throughput = \frac{Number\ of\ Delivered\ Packets * Packet\ Size * 8}{Total\ duration\ of\ simulation} \tag{23}$$

4.3 Results and Performance Analysis

This section illustrates the performance of the proposed technique. It is numerically analyzed comparatively with the classical cluster-based LEACH protocol.Results are illustrated based on the average results of 10 independent runs based on same randomized deployments of WSN(A, B, C and D).

Figures 1, 2, 3 and 4 indicate the respective round number for the death of the first node, death of 50 % of nodes and the death of last node in the network. The larger the first node dies the better optimization for the proposed technique achieved. This

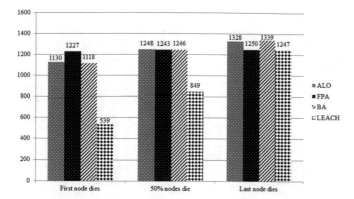

Fig. 1 Lifetime for WSN(A)

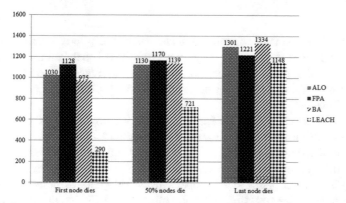

Fig. 2 Lifetime for WSN(B)

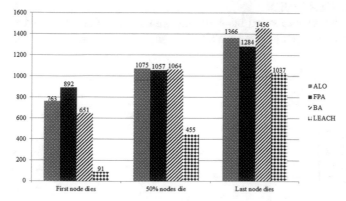

Fig. 3 Lifetime for WSN(C)

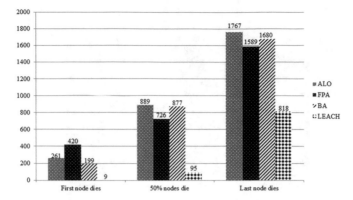

Fig. 4 Lifetime for WSN(D)

proves the efficiency of selecting the optimal cluster head based on the nodes' residual energy as well as the efficiency in cluster formation based on the intra-cluster distance between nodes that preserve the consumed energy. Figures 5, 6, 7 and 8 illustrate the number of the total of dead nodes across the WSN lifetime (*number of rounds*) for the four WSNs.

From all the previous WSN's lifetime analysis, it is obvious that the proposed natural inspired optimization model is greatly superior to LEACH concerning improving the lifetime of wireless sensor network. It is clear that nodes in WSN based on this model die in latter rounds. This proves that the proposed model achieves the objective in associating cluster nodes to the proper cluster head node with the minimum distance. Moreover to the selection of the best cluster head that has the highest available energy in each cluster. This implies to reserve the total network energy and hence extend the WSN lifetime.

The efficiency of natural inspired algorithms over LEACH caused by the better cluster formation of clusters in WSN. The selection of cluster head nodes is based

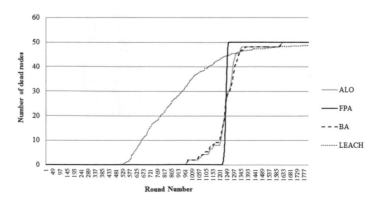

Fig. 5 Lifetime for WSN(A)

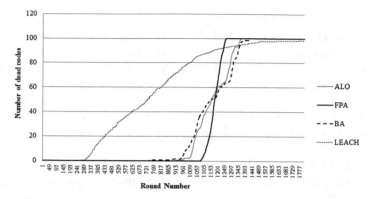

Fig. 6 Lifetime for WSN(B)

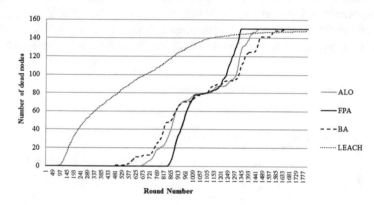

Fig. 7 Lifetime for WSN(C)

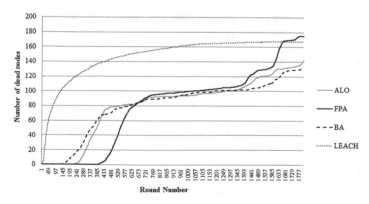

Fig. 8 Lifetime for WSN(D)

on choosing the nodes with the highest residual energy even if these nodes were previously selected as a cluster heads in previous rounds. Cluster formation of the proposed model associates nodes to each cluster based on the minimum distances between nodes and cluster head node for each cluster. This reduces the amount of consumed energy for communication and data transmission and hence increases the total available energy of nodes and WSNs. It saves the network energy by selecting the *CH* with highest energy no matter if it is have been previously selected or not and formatting clusters based on the minimum intra-cluster distances between nodes and *CH*.

Figures 9, 10, 11 and 12 illustrate the sum of the total available energy in WSN nodes (*residual energy*) across the WSN lifetime (*number of rounds*) for the four WSNs.

The comparative analysis of the proposed model comparatively with the LEACH protocol shows that LEACH has lower sum of total WSN's available energy across all the rounds (residual energy). The available WSN's total energy has drastically drained. This implies that the proposed natural inspired optimization model optimizes

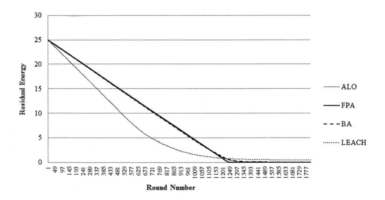

Fig. 9 Sum of total available energy for WSN(A)

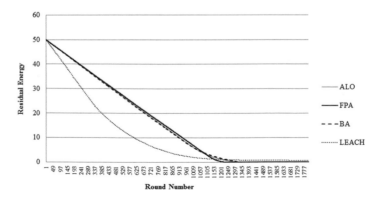

Fig. 10 Sum of total available energy for WSN(B)

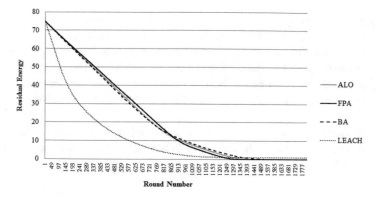

Fig. 11 Sum of total available energy for WSN(C)

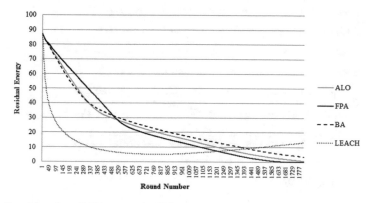

Fig. 12 Sum of total available energy for WSN(D)

and increases the availability of the total amount of energy of the WSN and hence, provides more power to extend the network lifetime.

The proposed optimization model superimpose based on the minimization of the intra-cluster distances between nodes and the associated cluster heads in each cluster. LEACH requires source nodes to send data directly to cluster heads. However, if the cluster head is far away from the source nodes, they might expend excessive energy in communication. Furthermore, LEACH requires cluster heads to send their aggregated data to the sink over a single-hop link. However, single-hop transmission may be quite expensive when the sink is far away from the cluster heads. LEACH also makes an assumption that all sensors have enough power to reach the sink if needed which might be infeasible for energy constrained sensor nodes.

In addition of overcoming the LEACH energy drawback, optimization of residual energy in WSN is a one of the main reasons to reach the stability intervals. The fixed numbers of nodes across rounds that preserve a constant or slightly variant energy level across the network preserve stable state with a little fluctuation. The proposed

technique reached global maximization of WSN lifetime based on applying optimal
selection of CH nodes which effectively preserve the global energy.

The numerical analysis and performance metrics conclude that the bat swarm
meta-heuristic proposed model generates a feasible scalable optimal energy-efficient
routing approach to optimize the WSN's lifetime. It overcomes the classical low-
energy adaptive clustering hierarchy algorithm; LEACH. This effectively extend the
WSNs lifetime.

Figures from 13, 14, 15 and 16 trace the number of received packets to the
Sink/Base Station node over the rounds of WSN's lifetime for the four WSNs used
(A, B, C and D).

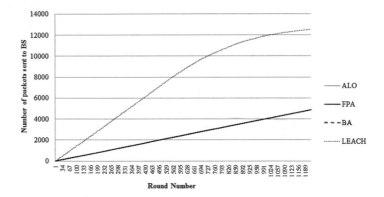

Fig. 13 Throughput for WSN(A)

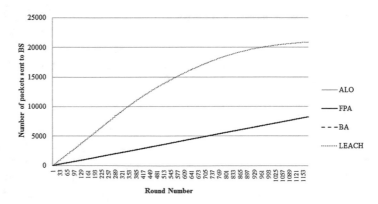

Fig. 14 Throughput for WSN(B)

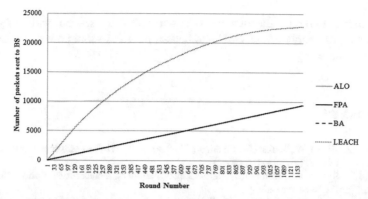

Fig. 15 Throughput for WSN(C)

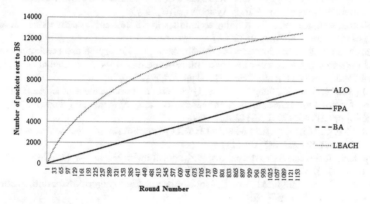

Fig. 16 Throughput for WSN(D)

5 Conclusion

A system for WSN life time maximization is proposed in this chapter based on the optimal selection of cluster heads and hierarchical routing. A set of recently proposed meta-heuristic optimizers were applied to find the optimal cluster heads maximizing cluster compactness. The proposed system is simulated on four different random network topologies and is assessed using different assessment indicators and is benchmark against LEACH as a common method for CH selection.

It is numerically proved from results that using the natural inspired proposed model generates a feasible scalable optimal energy-efficient routing approach to optimize the WSN's lifetime. It overcomes the classical low-energy adaptive clustering hierarchy algorithm; LEACH. The proposed model is greatly superior to LEACH concerning improving the lifetime of wireless sensor network as nodes in WSN based die in latter rounds. This proves that proposed model achieve the objective in associating cluster nodes to the proper cluster head node with the minimum distance. The selection of the best cluster head that has the highest available energy in each cluster implies to reserve the total network energy. It increases the availability of the total amount of energy of the WSN and hence, provides more power to extend the network lifetime.

References

1. Akyildiz, I.F., Su, W., Sankarasubramaniam, Y., Cayirci, E.: Wireless sensor networks: a survey. Comput. Netw. **38**(4), 393–422 (2002)
2. Hill, J.L.: System architecture for wireless sensor networks. Doctoral dissertation, University of California, Berkeley (2003)
3. Becker, M., Schaust, S., Wittmann, E.: Performance of routing protocols for real wireless sensor networks. In: Proceedings of the 10th International Symposium on Performance Evaluation of Computer and Telecommunication Systems (2007, July)
4. Al-Karaki, J.N., Kamal, A.E.: Routing techniques in wireless sensor networks: a survey. IEEE Wireless Commun. **11**(6), 6–28 (2004)
5. Heinzelman, W.R., Chandrakasan, A., Balakrishnan, H.: Energy-efficient communication protocol for wireless microsensor networks. In: Proceedings of the 33rd Annual Hawaii International Conference on System Sciences, 10 pp. IEEE (2000, January)
6. Xu, J., Jin, N., Lou, X., Peng, T., Zhou, Q., Chen, Y.: Improvement of LEACH protocol for WSN. In: 2012 9th International Conference on Fuzzy Systems and Knowledge Discovery (FSKD), pp. 2174–2177. IEEE (2012, May)
7. Abad, M.F.K., Jamali, M.A.J.: Modify LEACH algorithm for wireless sensor network. IJCSI Int. J. Comput. Sci. Iss. **8**(5) (2011)
8. Yang, X.S.: A new metaheuristic bat-inspired algorithm. In: Nature Inspired Cooperative Strategies for Optimization (NICSO 2010), pp. 65–74. Springer, Berlin (2010)
9. Yang, X.S., Karamanoglu, M., He, X.: Multi-objective flower algorithm for optimization. Proc. Comput. Sci. **18**, 861–868 (2013)
10. Yang, X.S.: Flower pollination algorithm for global optimization. In: Unconventional Computation and Natural Computation, pp. 240–249. Springer, Berlin (2012)
11. Pavlyukevich, I.: Lvy flights, non-local search and simulated annealing. J. Comput. Phys. **226**(2), 1830–1844 (2007)
12. Mirjalili, S.: The ant lion optimizer. Adv. Eng. Softw. **83**, 80–98 (2015)
13. Li, B., Zhang, X.: Research and improvement of LEACH protocol for wireless sensor network. Lect. Notes Inf. Technol. **25**, 48 (2012)
14. Bhadeshiya, J.R.: Improved performance of LEACH for WSN using precise number of cluster-head and better cluster-head selection. Int. J. Sci. Res. Dev. **1**(2) (2013)
15. Kaur, H., Seehra, A.: Performance evaluation of energy efficient clustering protocol for cluster head selection in wireless sensor network. Int. J. Peer to Peer Netw. (IJP2P) **5**(3), 1–5 (2014)
16. Anitha, R., Kamalakkannan, P.: Performance evaluation of energy efficient cluster based routing protocols in mobile wireless sensor networks. Int. J. Eng. Sci. Technol. **5**(6) (2013)
17. Oily Fossils Provide Clues to the Evolution of Flowers, Science Daily, 5 April 2001. http://www.sciencedaily.com/releases/2001/04/010403071438.htm. Last visited January 2015
18. Glover, B.J.: Understanding Flowers and Flowering: An Integrated Approach, vol. 277. Oxford University Press, Oxford (2007)
19. Pavlyukevich, I.: Lvy flights, non-local search and simulated annealing. J. Comput. Phys. **226**(2), 1830–1844 (2007)

Discrete Firefly Algorithm for Recruiting Task in a Swarm of Robots

Nunzia Palmieri and Salvatore Marano

Abstract In this chapter, we propose a Discrete Firefly Algorithm (DFA) for mine disarming tasks in an unknown area. Basically, a pheromone trail is used as indirect communication among the robots, and helps the swarm of robots to move in a grid area and explore different regions. Since a mine may need multiple robots to disarm, a coordination mechanism is necessary. In the proposed scenario, decision-making mechanism is distributed and the robots make the decision to move, balancing the exploration and exploitation, which help to allocate necessary robots to different regions in the area. The experiments were performed in a simulator, testing the scalability of the proposed DFA algorithm in terms of number of robots, number of mines and the dimension of grid. Control parameters inherent to DFA were tuned to test how they affect the solution of the problem.

Keywords Swarm intelligence · Swarm robotics · Firefly Algorithm · Nature-inspired algorithms

1 Introduction

In applications that do not allow human beings to access, multi-robot systems can play an important role to accomplish some tasks. Possible applications are exploration, search and rescue, monitoring, surveillance, cleaning. In order to successfully

N. Palmieri (✉) · S. Marano
Department of Computer Engineering, Modeling, Electronics,
and Systems Science, University of Calabria, Rende, CS, Italy
e-mail: n.palmieri@dimes.unical.it; n.palmieri@mdx.ac.uk

S. Marano
e-mail: marano@dimes.unical.it

N. Palmieri
School of Science and Technology, Middlesex University,
The Burroughs, London NW4 4BT, UK

© Springer International Publishing Switzerland 2016
X.-S. Yang (ed.), *Nature-Inspired Computation in Engineering*,
Studies in Computational Intelligence 637, DOI 10.1007/978-3-319-30235-5_7

133

perform, distributed coordination and cooperation mechanisms are desirable for multi-robot systems, especially, under a dynamic environment, to ensure robustness, flexibility, and reliability to the system [1].

Swarm Robotics is a new approach to the coordination of multi-robot systems that consists in a large number of simple robots, that are autonomous, not controlled centrally, capable of local communication, that behave like a team and not merely as single entities and operates based on some sense of biological inspiration. Swarm robotics and the related concept of swarm intelligence, is inspired by an understanding of the decentralized mechanisms that underlie the organization of natural swarms such as ants, bees, birds, fish, wolfs and even humans. Social insects provide one of the best-known examples of biological self organized behavior. By means of local and limited communication, they are able to accomplish impressive behavioral feats. The analysis of the social characteristics of insects and animals is essential in order to apply these findings to the design of multi-robot systems and development of similar behaviours in a coordination tasks [2].

A key component of this system is the communication between the agents in the group which is normally local, and guarantees the system to be scalable and robust. A plain set of rules at individual level can produce a large set of complex behaviors at the swarm level. The rules of controlling the individuals are abstracted from the cooperative behaviors in the nature swarm. The swarm is distributed and decentralized, and the system shows high efficiency, parallelism, scalability and robustness [3].

Various communication mechanisms are used for the communication of the swarm in order to accomplish the coordination and cooperation tasks. Many of them try inspiration from the nature; nature-inspired metaheuristics represent a class of heuristic methods inspired by natural behaviors that have ensured the survivability of animals, birds, insects by enabling them to find perfect solutions to almost all their problems (food search, breading, mating). A new trend in developing nature-inspired metaheuristics is to combine the algorithmic components from various metaheuristic aiming to improve the performance of the original techniques in solving NP hard problems. In our case we used a modified version of Ant Colony Optimization [4] for the exploration task and a discrete version of Firefly Algorithm [5] to disarming task.

Basically, in this work, we consider autonomous exploration and search in an unknown environment. The goal is to explore in a minimum amount of time the environment, while searching for the mines that are disseminated in the area.

The environment is not known beforehand and it has to be fully explored; at the same time, the environment must be analyzed in order to detect and disarm some mines disseminated in the area.

More specifically, we focus on multi-objective exploration and search. On one hand the robots explore in independent manner the environment, distributing in the area trying to minimize the overall exploring time avoiding passing in the same region more time. On the other hand, when a mine is detected, multiple robots are needed to work together to accomplish the task so it is necessary a coordinated strategy among involved robots.

To address the social issues in the exploration, the algorithm is based on our recently published heuristic model, inspired by the foraging behavior of colonies of ant in nature [7]. During the exploration, the robots deposit in the terrain a quantity of pheromone that is perceived by other robots that make a decision based on the minimum amount of pheromone in the neighborhood. When they find a mine, a robot becomes a firefly and trying to recruit (attract) other robots according to a novel bio inspired algorithm called Firefly Algorithm [5, 6]. The information about the location of the mines is passed using broadcast communication in order to inform other robots, in the wireless range, about the mine. If a robot receives more requests of recruitment, it evaluates the quality of the target based on the minimum distance [8]. The strategy is able to adapt the current system dynamics if the number of robots or the environment structure or both change.

The algorithm has been previously validated, and this paper presents the analysis of the value of the proposed Firefly Algorithm parameters in a simulated environment. The simulations were performed for various sets of variables, including number of robots; number of mines and dimension of search area. Control parameters inherent to the DFA were tuned to test how they affect the time to complete the overall task (exploring and disarming).

The rest of this paper is organized as follows. In Sect. 2 we briefly revisit approaches related to the problem; in the Sect. 3, we describe the Firefly Algorithm; in Sect. 4 we analyze the problem and the proposed method. In Sect. 5, we report the simulation results tuning the control parameters of the algorithm, including a statistical considerations and finally, we provide concluding remarks in Sect. 6.

2 Related Work

Nature inspired metaheuristics, relying on concepts and search strategies inspired from nature, have emerged as a promising type of stochastic algorithms used for efficiently solving optimization problems.

For sharing information and accomplishing the tasks there are, basically, three ways of information sharing in the swarm: direct communication (wireless, GPS), communication through environment (stigmergy) and sensing. More than one type of interaction can be used in one swarm, for instance, each robot senses the environment and communicates with their neighbor. Balch [9] discussed the influences of three types of communications on the swarm performance and Tan and Zheng [2] presented an accurate analysis for the different types of communication and the impact in a behavior of swarm.

Within the context of swarm robotics, most works on cooperative tasks are based on biologically behavior and indirect stigmergic communication (rather than on local information, which can be applied to systems related to GPS, maps, and wireless communications).

The self-organizing properties of animal swarms such as insects have been studied for better understanding of the underlying concept of decentralized decision-making in nature, but they also gave a new approach in applications to multi-agent systems engineering and robotics. Bio-inspired approaches have been proposed for multi-robot division of labor in applications such as exploration and path formation, or cooperative transport and prey retrieval.

Within the context of swarm robotics, most works on cooperative tasks are based on social behavior like Ant Colony Optimization. The principle is simple: ants deposit a pheromone trail on the path they take during travel. Using this trail, they are able to navigate toward their nest or food and communicate with their peers. More specifically, ants employ an indirect recruitment strategy by accumulating pheromone trails. When a trail gets strong enough, other ants are attracted to it and will follow this trail toward a food destination. The more ants follow a trail, the more pheromone is accumulated and in turn, the trail becomes more attractive for being followed. Pheromone in swarm robotics can be viewed as a mechanism for inter-robot communication that can help reduce the complexity of individual agents [4, 10].

More strategies are based on these principles: Labella et al. [11] have analyzed the behavior of a group of robots involved in an object retrieval task, trying inspiration by a model of ants' foraging.

Other authors experiment with chemical pheromone traces, e.g. using alcohol like in Fujisawa et al. [12] or using a special phosphorescent glowing paint such as has presented by Mayet et al. [13].

An improved form of pheromone communication method called virtual pheromone was used by Payton et al. [14] to employ simple communication and coordination to achieve large scale results in the areas of surveillance, reconnaissance, hazard detection, and path finding. De Rango and Palmieri [7] used an Ant-based strategies incorporating both attraction and repulsion features.

Particle Swarm Optimization is another biologically-inspired algorithm [15] motivated by a social analogy, such as flocking, herding, and schooling behavior in animal populations and have received much attention in recent year.

Pugh and Martinoli [16] proposed an adapted version of Particle Swarm Optimization learning algorithm to distribute unsupervisioned robotic learning in groups of robots with only local information. More applications in swarm robotic use this strategy to coordinate the swarm in a collective search [17] or task allocation like in Meng and Gan [18].

Another very good example of natural swarm intelligence is honeybee foraging, because the group of foraging bees is not controlled by a central decision-making unit and because the collective decisions of colonies in varying environments were found to act in an intelligent way. In honeybees, foraging decisions are based on such local cues, which are exploited by individual forager and receiver bees. Based on these cues and on communication transferred through bee dances [19], one worker can affect the behavior of other nearby workers. This way behavioral feedback loops emerge, which lead to a self-organized regulation of workforce and foraging decisions. Jevetic et al. [20] proposed a distributed bees algorithm (DBA) for task allocation in a 2-D arena.

Explicit communication is a type of communication in which the robots directly pass messages to each other using for example wireless communication. In this case, a communication protocol and an efficient routing algorithm are important for a coordination of the swarm of mobile robots and for a robustness of the total system. In this case ACO routing algorithms can be used to solve routing problem and load balancing in the network. De Rango et al. [21–24] proposed novel routing algorithms based on Swarm Intelligence able to satisfy multiple metrics for a multi-objective optimization like end to end delay, load balancing and energy savings.

Our approach combines an implicit communication among the robots during the exploration task and kind of explicit communication among the robots during the recruiting task. During the exploration, the robots mark the crossed cell through the scent that can be detected by the other robots; the robots choose the cell that has the lowest quantity of substance to allow the exploration of the unvisited cells in order to cover the overall area in less time. After a while, the concentration of pheromone decreases due to the evaporation and diffusion associated with the distance and with the time; in this way we can allow continuous coverage of an area via implicit coordination. The other robots, through proper sensors, smell the scent in the environment and move in the direction with a minimum amount of pheromone that corresponds to an area less occupied and probably an unexplored area [7]. On the other hand, in order to deactivate the mines, the first robot that detects a mine (firefly) in a cell, tries to attract the other robots according to a novel bio-inspired approach called Firefly algorithm (FA) summarized in the next section.

3   The Firefly Algorithm

Firefly Algorithm (FA) is a nature-inspired stochastic global optimization method that was developed by Yang [5, 6]. The FA tries to mimic the flashing behavior of swarms of fireflies. In the FA algorithm, the two important issues are the variation of light intensity and the formulation of attractiveness. The brightness of a firefly is determined by the landscape of the object function. Attractiveness is proportional to brightness and, thus, for any two flashing fireflies, the less bright one move towards the brighter one. The light intensity decays with the square of the distance, the fireflies have limited visibility to other fireflies. This plays an important role in the communication of the fireflies and the attractiveness, which may be impaired by the distance. Some simplifications are assumed such as:

- it is assumed that all fireflies are unisex so they will be attracted to each other regardless of their sex.
- The attractiveness is proportional to their brightness and they both decrease as the distance increases.
- In the case of no existence of no brighter firefly on then, the fireflies will move randomly.
- The brightness of firefly is affected by its fitness.

The distance between any two fireflies i and j, at positions x_i and x_j, respectively, can be defined as the Euclidean distance as follows:

$$r_{ij} = \|x_i - x_j\| = \sqrt{\sum_{k=1}^{D}(x_{i,k} - x_{j,k})^2} \tag{1}$$

As a firefly's attractiveness is proportional to the light intensity seen by adjacent fireflies, the variation of attractiveness β with the distance r is defined as followed:

$$\beta = \beta_0 e^{-\gamma r_{ij}^2} \tag{2}$$

The movement of a firefly i is attracted to another more attractive firefly j is determined by:

$$x_i^{t+1} = x_i^t + \beta_o e^{-\gamma r_{ij}^2}\left(x_j - x_i\right) + \alpha(\sigma - \frac{1}{2}) \tag{3}$$

The Firefly Algorithm has been proved very efficient and it has three key advantages [25]:

- Automatic subdivision of the whole population into subgroups so that each subgroup can swarm around a local mode. Among all the local modes, there exists the global optimality. Therefore, FA can deal with multimodal optimization naturally.
- FA has the novel attraction mechanism among its multiple agents and this attraction can speed up the convergence. The attractiveness term is nonlinear, and thus may be richer in terms of dynamical characteristics.
- FA can include PSO, DE, and SA as its special cases. Therefore, FA can efficiently deal with a diverse range of optimization problems.

4   Problem Statement

We consider an environment assuming that it is discretized into equally spaced cells that contains a certain number of mines. Robots can move among cells and they can have just local information about robots (neighbors) or regions to explore (neighbor cells) in order to provide a scalable strategy. They are assumed to be able to move in the grid; there are N robots (R_1; $:::$; R_N) placed initially in random positions on a unit grid. The considered scenario is presented under this assumption.

1. The robots are equipped with proper sensors that are able to deposit and smell the chemical substances (pheromone) leaved by the other robots in order to guide the swarm in the exploration task.
2. They make a probabilistic decision based on amount of pheromone in the cells.
3. They are equipped with proper sensor to detect the mines.

4. The robots are equipped with wireless module, indeed when a robot detects a mine, it becomes a firefly and tries to attract other robots sending messages via broadcast communication to robots in the wireless range.
5. The robots, that receive messages by different robots (fireflies), evaluate the light of fireflies and choose the best firefly (in our case the fireflies at minimum distance) and move toward firefly according to the Firefly Algorithm.

It is assumed that they are able to disarm a mine cooperatively. In this paper we investigated the recruiting strategy in order to lead the robots in the positions of the mines in the area and we did not consider properly the disarming task.

In our work, the map of the environment will be generated as a set of grid cells. The robots know only the adjacent cells and during the exploration they spray a scent (pheromone) into the cells to support the navigation of the others. In the algorithm, the robots decide the direction of the movement relying on a probabilistic law inherited by swarm intelligence and swarm robotics techniques. The scent evaporates not only due to diffusion effects in the time, but also in the space according to the distance this allows a higher concentration of scent in the cell where the robot is moving and a lower concentration depending on the distance [7].

Each robot (r), according to uniform distribution, stores in a grid-based map that is represented by a matrix M (dimension $m \times n$). It is assumed that each robot in a cell $(i, j) \in M$ can move just in the neighbor cells through discrete movements. Let z be the number of mines on a set MS to distribute on the grid area according to an uniform distribution. The MS set is characterized by the coordinates of the mines. The objective is to detect and disarm all mines and discovery all cells in the grid (this last condition assures that all mines could be correctly detected in an unknown area). Let t_e be the time necessary for a robot to visit a cell, and let t_d be the time necessary to disarm a mine once it has been detected. It is assumed that a fixed number of robots ($rd_{\min}$) are necessary to disarm a mine stored in a cell $(i, j) \in MS$. This means that for the exploration task, the robots can be distributed among the area because each robot independently explores all cells, whereas for the mine detection, more robots need to be recruited in the same cell in order to perform the task. $M(i, j)\_a$ is a variable representing the number of robots (accesses) that passed through the cell (i, j). We define a bi-objective function as both the time to detect and the disarming the mine through the exploration on the overall grid.

$$\min \sum t_e \quad \text{and} \quad \min \sum_{i=1}^{z} t_{d,i} \tag{4a}$$

subject to

$$M(i, j)\_a \geq 1 \quad i = 1 \ldots m; \quad j = 1 \ldots n \text{ with } (i, j) \in M$$
$$M(i, j)\_a \geq rd_{\min} \text{ with } (i, j) \in MS$$

This is a bi-objective optimization problem and its solutions will result in a Pareto front. However, in order to solve this problem more effectively, for simplicity, we will combine these two objectives to form a single objective optimization problem so as to minimize the overall total time as follows:

$$\min T_{tot} = \min \left(\sum t_e + \sum_{i=1}^{z} t_{d,i} \right) \tag{4b}$$

subject to

$$M(i, j)\_a \geq 1 \quad i = 1 \ldots m; \quad j = 1 \ldots n \text{ with } (i, j) \epsilon M$$
$$M(i, j)\_a \geq r d_{min} \text{ with } (i, j) \epsilon M S$$

4.1 Exploration Task

The overall task can be divided into two sub tasks: exploration and disarming task. For the exploration we use a strategy inherit swarm intelligent trying inspiration by a modified version of Ant Colony Optimization. The law used by the robots to choose the cells during the movement, according with De Rango and Palmieri [7], is presented below. We consider a robot in a cell s and it will attribute to the set of next cells v_i following a probability as:

$$p(v_i | s) = \frac{[\tau_{v_i,t}]^\varphi \cdot [\eta_{v_i,t}]^\vartheta}{\sum_{i \in N(s)} [\tau_{v_i,t}]^\varphi \cdot [\eta_{v_i,t}]^\vartheta}, \quad \forall v_i \in N(s) \tag{5}$$

where $p(v_i|s)$ represents the probability that the robot, that is in the cell s, chooses the cell v_i; $N(s)$ is the set of neighbors to the cell v_i; $\tau_{vi,t}$ is the amount of pheromone in the cell vi; η_{vi}, t is the heuristic parameter introduced to make the model more realistic. In addition, φ and θ are two parameters which affect respectively the pheromone and heuristic values (Table 1). Taking into account the spatial dispersion of the pheromone and the temporal dispersion the amount of pheromone in the cell v where the robot will move during the exploration is:

$$\tau_{v,t+1}(d) = t_{v,t} + t_v(d) \tag{6}$$

In order to explore different areas of the environment, the robots choose the cell with a minimum amount of pheromone, corresponding to cell that probably is less frequented and therefore not explored cell. The chosen cell will be selected according with Eq. (5).

$$v_{next} = \min [p(v_i|s)] \quad \forall v_i \in N(s) \tag{7}$$

Table 1 Symbols adopted in the problem

Symbols	Meaning
r_{ij}	Euclidean distance
β_0	Attractiveness coefficient
γ	Absorption coefficient
σ	Random number
α	Randomization parameter
t_e	Time to visit a cell
$t_{d,i}$	Time to disarm the mine i
τ_{vi}, t	Pheromone in cell v_i at the time t
$\eta_{vi,t}$	Heuristic parameter in the cell v_i at time t
φ	Pheromone parameter
θ	Heuristic parameter
$p(v_i|s)$	Probability that a robot r, that stores in the cell s, chooses the cell v_i
z	Number of mines
T_{tot}	Total time to complete the task (exploration and disarming)
$M(i, j)\_a$	Number of accesses in the cell (i, j)

4.2 Proposed Discrete Firefly Algorithm for Disarming Task

When a robot finds a mine, during the exploration task, it becomes a recruiter (firefly) of the other robots in order to disarm the mine and it tries to attract the other robots based on the mine position.

In this case, the robots are assumed to have transmitters and receivers, using which they can communicate messages to each other. The messages are mostly coordinate positions of the detected mines. However, the robots are assumed to be able to broadcast messages in their wireless range; in this way, a robot can transmits its position only to its neighbors directly and there is not propagation of the messages (one hop communication).

The original version of FA is applied in the continuous space [5], and cannot be applied directly to tackle discrete problem, so we modified the algorithm in order to fit with our problem. In our case, a robot can move in a discrete space because it can go just in the contiguous cells step-by-step. This means that when a robot perceives, in its wireless range, the presence of a firefly (the recruiter robot) and it is in a cell with coordinates x_i and y_i, it can move according with the FA attraction rules such as expressed below:

$$\begin{cases} x_i^{t+1} = x_i^t + \beta_o e^{-\gamma r_{ij}^2}\left(x_j - x_i\right) + \alpha\left(\sigma - \frac{1}{2}\right) \\ y_i^{t+1} = y_i^t + \beta_o e^{-\gamma r_{ij}^2}\left(y_j - y_i\right) + \alpha(\sigma - \frac{1}{2}) \end{cases} \tag{8}$$

Fig. 1 Two robots during
the exploration receive more
than one recruiting calls
because they are in an
overlapped area. The DFA
tries to coordinate the robots
in the disarming task
avoiding redundancy

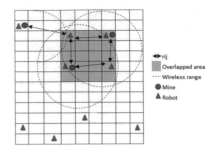

where x_j and y_j represent the coordinates of detected mine translated in terms of row
and column of the matrix area, r_{ij} is the Euclidean distance between mine (or firefly)
according to the Eq. (1) and the robot that moves towards the mine. The robots
movements are conditioned by mine (firefly) position, in the second term of the
formula, and it depends on attractiveness of the firefly such as expressed in Eq. (2)
and by a random component in the third term of Eq. (8).The coefficient α is a
randomization parameter determined by a problem of interest. The σ coefficient is
a random number generator uniformly distributed in the space [0, 1] and it is useful
to avoid that more robots go towards the same mine if more robots are recruited by
the same firefly and enabling to the algorithm to jump out of any local optimum
(Fig. 1). In order to modify the FA to a discrete version, the robots movements have
been considered through three possible value updates for each coordinates: $\{-1, 0, 1\}$
according to the following condition:

$$\begin{cases} x_i^{t+1} = x_i^t + 1 & if \left[\beta_o e^{-\gamma r_{ij}^2}\left(x_j - x_i\right) + \alpha\left(\sigma - \frac{1}{2}\right) > 0\right] \\ x_i^{t+1} = x_i^t - 1 & if \left[\beta_o e^{-\gamma r_{ij}^2}\left(x_j - x_i\right) + \alpha\left(\sigma - \frac{1}{2}\right) < 0\right] \\ x_i^{t+1} = x_i^t + 0 & if \left[\beta_o e^{-\gamma r_{ij}^2}\left(x_j - x_i\right) + \alpha\left(\sigma - \frac{1}{2}\right) = 0\right] \end{cases} \qquad (9)$$

$$\begin{cases} y_i^{t+1} = y_i^t + 1 & if \left[\beta_o e^{-\gamma r_{ij}^2}\left(y_j - y_i\right) + \alpha\left(\sigma - \frac{1}{2}\right) > 0\right] \\ y_i^{t+1} = y_i^t - 1 & if \left[\beta_o e^{-\gamma r_{ij}^2}\left(y_j - y_i\right) + \alpha\left(\sigma - \frac{1}{2}\right) < 0\right] \\ y_i^{t+1} = y_i^t + 0 & if \left[\beta_o e^{-\gamma r_{ij}^2}\left(y_j - y_i\right) + \alpha\left(\sigma - \frac{1}{2}\right) = 0\right] \end{cases} \qquad (10)$$

A robot r, which stores in the cell (x_i, y_i) as depicted in Fig. 2, can move in eight
possible cells according with the three possible values attributed to x_i and y_i.

For example if the result of the Eqs. (9–10) is $(-1, 1)$ the robot will move in the
cell $(x_i - 1, y_i + 1)$.

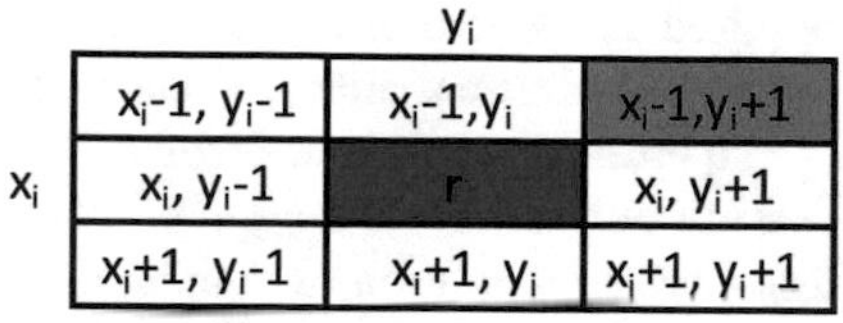

Fig. 2 Possible movement for a robot r stores in a cell (x_i, y_i)

	y_i		
	x_i-1, y_i-1	x_i-1,y_i	x_i-1,y_i+1
x_i	x_i, y_i-1	r	x_i, y_i+1
	x_i+1, y_i-1	x_i+1, y_i	x_i+1, y_i+1

In the described problem, the firefly algorithm is executed as follows:

- Step 1: Get the list of the detected mines (fireflies) and initialize algorithm's parameters: z number of fireflies, the attractiveness coefficient β_0, the light absorption coefficient γ, randomization parameter α.
- Step 2: Get the list of the robots in the wireless range of the fireflies.
- Step 3: For each robot calculate the distance r_{ij} from the fireflies in its range using the Euclidian distance.
- Step 4: For each robot find the firefly at minimum distance (the best firefly) and try to move the robots from their locations to the location of the best firefly according to the Eqs. (9)–(10).
- Step 6: Terminate if all detected mines are disarmed.

These steps are executed when the robots are recruited by others, indeed when no fireflies are detected or if the new location of the robots is outside of the wireless range of the fireflies, the robots explore in independent manner the area according the Eq. (5). This happens because the nature of the problem is bi-objective and the robots have to balance the two tasks.

5 Parameter Settings

A suitable setting for the DFA parameters values is needed for the proposed mechanism in order to work effectively. We have conducted experiments to find out the most suitable parameter values.

In this section, we evaluate the parameters, focusing on the overall time that is defined as the time needed for the robots to cover the area and disarm all mines disseminated in the area. We considered the overall time because the strategy of recruitment affects significantly the exploration strategy and finally the overall time.

To highlight the performance benefits, we used random positions of the mines and the robots in the area, varying the number of robots operating in the 2-D arena, the size of grid map and the number of robots needed to disarm a mine, in order to study the performance and scalability of the proposed DFA algorithm.

We considered three scenarios: the first scenario is represented by a grid map 20×20 with 3 mines to discover and disarm; the second scenario is represented by a grid map 30×30 with 5 mines to discover and disarm. The third scenario considered a grid map 40×40 with 10 mines to discover and disarm. In the first scenario are required 2 robots to disarm a mine; in the second and third scenario we considered 4 robots to disarm a mine, increasing the complexity of the task.

Experiment setup was created to check the influence of the control parameters in solutions, evaluating the time to complete all task (exploring and disarming task) measured in term of number of iterations. Each of the numerical experiments was repeated 50 times considering the following values: the absorption coefficient ($\gamma = \frac{1}{L}; \frac{0.5}{L}; \frac{2}{L}; \frac{1}{\sqrt{L}}$), where L is max $\{m, n\}$ and m and n are the number of rows and columns of the matrix M that represents the size of the grid area; randomization parameter ($\alpha = 0.1; 0.2; 0.5$) and the attraction coefficient ($\beta_0 = 1; 0.2; 0.5$).

The value of this parameters is important, expecially, when the number of robots to coordinate is low. Figure 3a–c show the performance relative to the attractiveness. It is possible to see that there is no significant difference using different value of the β_0 when the task in not complicated (Fig. 3a). When the complexity of the task increases (Fig. 3b) an high coefficient of attraction $\beta_0 (\beta_0 = 1)$ influences negatively the performances when the number of robots is low and tend to be comparable when the number of robots increases. This is happens because an high value of the attractiveness means that the weight of the attraction in the Eq. (8) increases and it is possible that more robots, in an overlapped region, go towards the same mine, creating a not necessary redundancy increasing the time to complete the task. When the number of robot increases, the results are comparable.

Figure 4a–c show the trend of the randomization parameter. For the low number of the robots it is necessary balancing the attractiveness and randomization parameter in order to minimize the overall time, so the value of this parameter is important especially when the complexity of task increases in term of number of mines and number of robots needed to disarm a mine. It is reasonable to expect that by increasing the number of robots the efficiency of swarm improves and the values of parameter do not influence al lot the total performance.

In Fig. 5a–c are plotted the performance of the algorithm considering the absorption coefficient γ; its value is important when the complexity of the task increase as shown in Fig. 5c. In this case a high value of $\gamma (\gamma = 1/\sqrt{L})$ influences negatively the attractiveness of the firefly making the firefly less bright to recruit the other robots increasing the time to complete all tasks.

5.1 Solution Quality Analysis

In order to determine which of the related parameters are statistically better for the problem we have also considered the p-values of two-samples Student t-tests. The t-tests is used to analyze the relationships between the results obtained from different simulation groups. The experiments are referred to the first scenario (grid 20×20, 3 mines to disarm) second scenario (grid 30×30, 5 mines to disarm) and third scenario (grid 40×40, 10 mines to disarm). The traditionally accepted p-value for something

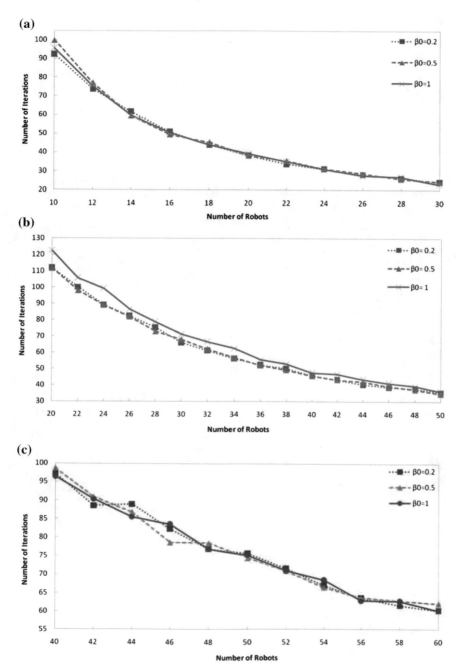

Fig. 3 Effect of the control parameter β_0 on the total time to complete the task measured by the number of iterations, **a** grid 20×20, 3 mines to disarme; **b** grid 30×30, 5 mines to disarme; **c** grid 40×40, 10 mines to disarme

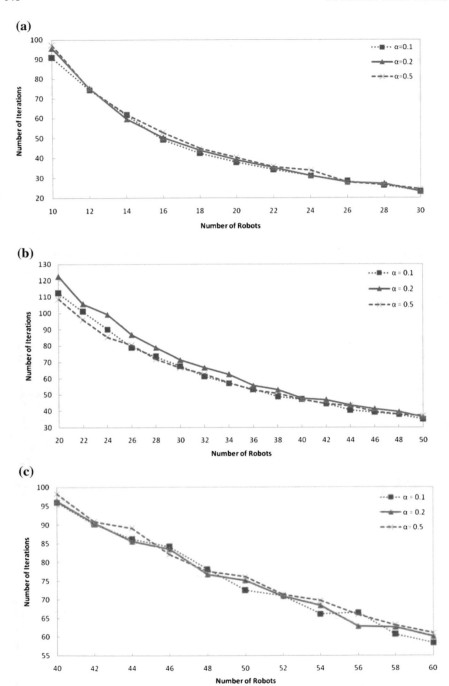

Fig. 4 Effect of the control parameter α on the total time to complete the task measured by the number of iterations, **a** grid 20×20, 3 mines to disarm; **b** grid 30×30, 5 mines to disarme; **c** grid 40×40, 10 mines to disarme

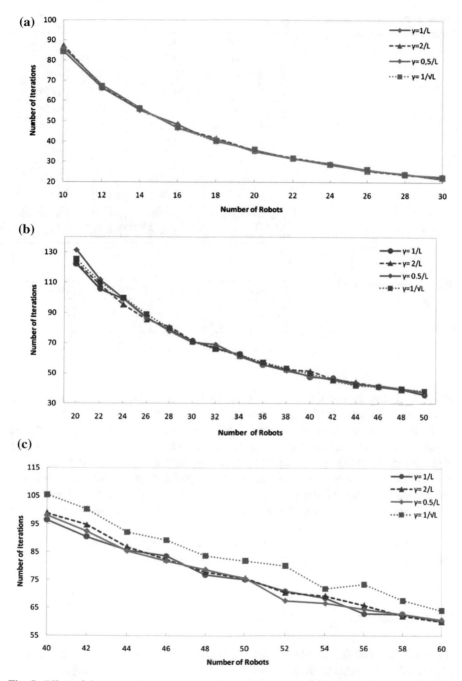

Fig. 5 Effect of the control parameter γ on the total time to complete the task measured by the number of iterations, **a** grid 20×20, 3 mines to disarme; **b** grid 30×30, 5 mines to disarme; **c** grid 40×40, 10 mines to disarme

Table 2 Results of p-values in the t-test for the DFA

Parameters	Scenario 1	Scenario 2	Scenario 3
$\beta_0 = 0.2$ and $\beta_0 = 0.5$	0.2529	0.6847	0.9335
$\beta_0 = 0.2$ and $\beta_0 = 1$	0.5150	**1.55E-05**	0.9524
$\beta_0 = 0.5$ and $\beta_0 = 1$	0.1882	**1.68E-05**	0.9060
$\alpha = 0.1$ and $\alpha = 0.2$	0.1427	**166E-05**	0.6385
$\alpha = 0.1$ and $\alpha = 0.5$	**0.0091**	0.5338	**0.0358**
$\alpha = 0.2$ and $\alpha = 0.5$	**0.0103**	**0.0008**	**0.0204**
$\gamma = \frac{1}{L}$ and $\gamma = \frac{2}{L}$	0.5348	0.3323	**0.0839**
$\gamma = \frac{1}{L}$ and $\gamma = \frac{0.5}{L}$	0.9325	0.1790	0.8886
$\gamma = \frac{1}{L}$ and $\gamma = \frac{1}{\sqrt{L}}$	0.7520	**0.0433**	**2.42E-06**
$\gamma = \frac{2}{L}$ and $\gamma = \frac{1}{\sqrt{L}}$	0.4058	0.3216	**6.95E-07**
$\gamma = \frac{0.5}{L}$ and $\gamma = \frac{1}{\sqrt{L}}$	0.6979	0.8691	**3.14E-04**
$\gamma = \frac{2}{L}$ and $\gamma = \frac{0.5}{L}$	0.3522	0.4459	0.0518

Table 3 Best parameters setting

Parameters	Scenario 1	Scenario 2	Scenario 3
α	0.1	0.5	0.2
β	0.2	0.5	0.5
γ	$\frac{1}{L}$	$\frac{1}{L}$	$\frac{1}{L}$

to be significant is p $<$ 0.05. In this case, there is an evidence that the means are significantly different at the significance level reported by the p-value.

Table 2 shows the p-value obtained from the t-tests using all above simulation results by considering each parameter for all considered scenario. In the table are highlighted in bold the p-value more interesting. In this case, there is a difference about the value assumed to the parameter of the problem, so in this case it is better choose the value that influence positively the time. In the other cases, although there is a difference about the time, there is not statistically significant, so it is possible to choose any value of the considered parameter.

To summarize the observations of the parameter settings we described in Table 3 the final best parameters considering the different Scenarios.

6 Conclusions

Swarm intelligence based algorithms are very efficient in solving a wide range of optimization problems in diverse applications in science and engineering.

Firefly Algorithm (FA) is a new swarm intelligence metaheuristic and it has been proven to be effective methods for solving some hard optimization problems. In this chapter, its application for a recruiting task in a swarm of mobile robots is investigated.

One key issue is that all metaheuristics have algorithm-dependent parameters, and the values of these parameters will largely influence the performance of an algorithm. Our experiments, through simulation, showed that the control of parameters provide a mechanism to adjust the robot behavior depending on the dimension of robots (swarm) and the complexity of task.

Results, from the tests, show that the values of the parameters are important when the complexity of tasks increases in terms of dimension of the area and the number of mines disseminated in the area. In particular it is better in this case to balance the attraction of the fireflies (mines) and the random movement in order to distribute better the robots in the area and avoid any redundancy in any region that involve in an increase of the time to complete all tasks.

Future work will include the analysis of the effect of these parameters considering other constraints like the battery that can pose a trade-off between efficiency of the motion and energy utilized. We also will consider the continuous movement of the robots in the area of interest. In addition, we will consider the impact of these parameters introducing obstacles in the area and dropping wireless connection.

References

1. Bellingham, J.G., Godin, M.: Robotics in remote and hostile environments. Science **318**, 1098–1102 (2007)
2. Tan, Y., Zheng, Z.-Y.: Research advance in swarm robotics. Defence Technol. **9**(1), 18–39 (2013)
3. Mohan, Y., Ponnambalam, S.: An extensive review of research in swarm robotics. In: Nature and Biologically Inspired Computing. NaBIC, Word Congress (2009)
4. Dorigo, M., Stutzle, T.: Ant Colony Optimization, MIT Press. ISBN 0-262-04219-3 (2004)
5. Yang, X.S.: Firefly algorithms for multimodal optimization. Lect. Notes Comput. Sci. **5792**, 169–178 (2009)
6. Yang, X.S.: Firefly algorithm, stochastic test functions and designoptimisation. Int. J. Bio-Inspir. Comput. **2**(2), 78–84 (2010)
7. De Rango, F., Palmieri, N.: A swarm-based robot team coordination protocol for mine detection and unknown space discovery. In: Proceedings of the 8th International Conference on Wireless Communications and Mobile Computing, IWCMC, Limassol (2012)
8. De Rango, F., Palmieri, N., Yang, X.S., Marano, S.: Bio-inspired exploring and recruiting tasks in a team of distributed robots over mined regions. In: Proceedings of the International Symposium on Performance Evaluation of Computer and Telecommunication System, Chicago (2015)
9. Balch, T.: Communication, diversity and learning: cornerstones of swarm behaviour. In: Swarm Robotics, Lecture Notes in Computer Science, vol. 3342, p. 21e30. Springer (2005)
10. Ducatelle, F., Di Caro, G.A., Pinciroli, C., Gambardella, L.M.: Selforganized cooperation between robotic swarms. Swarm Intell. **5**(2), 73–96 (2011)
11. Labella, T.H., Dorigo, M., Deneubourg, J.L.: Division of labour in a group inspired by ants' foraging behaviour. ACM Transactionds Auton. Adapt. Syst. **l**(1), 4–25 (2006)
12. Fujisawa, R., Dobata, S., Kubota, D., Imamura, H., Matsuno, F.: Dependency by concentration of pheromone trail for multiple robots. In: Proceedings of ANTS 2008, 6th International

Workshop on Ant Algorithms and Swarm Intelligence, volume 4217 of LNCS, pp. 283–290. Springer (2008)

13. Mayet, R., Roberz, J., Schmickl, T., Crailsheim, K.: Antbots: a feasible visual emulation of pheromone trails for swarm robots. In: Proceedings of the 7th International Conference on Swarm Intelligence (ANTS), pp. 84–94 (2010)

14. Payton, D., Daily, M., Estowski, R., Howard, M., Lee, C.: Pheromone robotics. Auton. Robots **11**(3), 319–324 (2001)

15. Kennedy, J., Eberhart, R.: Particle swarm optimization. Process. IEEE Int. Conf. Neural Netw. **4**, 1942–1948 (1995)

16. Pugh, J., Martinoli A.: Multi-robot learning with particle swarm optimization. Proceedings of Fifth International JointConference on Autonomous Robots and Multirobot Systems, pp. 441–448. Japan (2006)

17. Hereford, J.M., Siebold, M., Nichols, S.: Using the particle swarm optimization algorithm for robotic search applications. IEEE Swarm Intell. Symp. (2007)

18. Meng, Y., Gan, J.: A distributed swarm intelligence based algorithm for a cooperative multi-robot construction task. IEEE Swarm Intell. Symp. (2008)

19. Seeley, T.D., Camazine, S., Sneyd, J.: Collective decisionmaking in honey bees: how colonies choose among nectar sources. Behav. Ecol. Sociobiol. **28**(4), 277–290 (1991)

20. Jevtic, A., Gutiérrez, A., Andina, D., Jamshidi, M.: Distributed bees algortithm for task allocation in swarm of robots. IEEE Syst. J. **6**, 2 (2012)

21. De Rango, F. Tropea, M.: Swarm intelligence based energy saving and load balancing in wireless ad hoc networks. In: Proceedings of the 2009 workshop on Bio-inspired algorithms for distributes systems, pp. 77–84. NY (2009)

22. De Rango, F. Tropea, M., Provato, A., Santamaria, A.F., Marano, S.: Minimum hop count and load balancing metrics based on ant behaviour over hap mesh. In: Proceedings of the Global Teleccomunications Conference, New Orleans (2008)

23. De Rango, F., Tropea, M.: Energy saving and load balancing in wireless ad hoc networks trough ant-based routing. In: Proceedings of the Internation Symposium on Performance Evaluation of Computer and Telecommunication Systems (2009)

24. De Rango, F., Tropea, M., Provato, A., Santamaria, A.F., Marano, S.: Multi-constraints routing algorithm based on swarm intelligence over high altitude platforms. In: Proceedings of the International Workshop on Nature Inspired Cooperative Strategies for Optimization, pp. 409–418. Acireale (2007)

25. Fister, I., Yang, X.S., Fister, D., Fister Jr., I.: Firefly algorithm: a brief review of the expanding literature. In: Cuckoo Search and Firefly Algorithm, pp. 347–360. Springer, NY (2014)

Nature-Inspired Swarm Intelligence for Data Fitting in Reverse Engineering: Recent Advances and Future Trends

Andrés Iglesias and Akemi Gálvez

Abstract This chapter discusses the very important issue of data fitting with curves and surfaces in reverse engineering. In this problem, given a (usually massive) cloud of (generally noisy) data points, the goal is to approximate its underlying structure through a parametric free-form curve or surface. The process begins with the conversion of this problem from the purely geometric approach based on the point cloud to the mathematical formulation of this problem as a nonlinear continuous optimization problem. Furthermore, this problem is generally multimodal and of a very high dimension, and exhibits challenging features such as noisy data and/or irregular sampling, making it hard (if not impossible) to be properly addressed by traditional mathematical optimization techniques. In this chapter, we claim that this curve/surface data fitting problem can be successfully addressed by following a nature-inspired computational intelligence approach. To focus our discussion, we consider three recent nature-inspired swarm intelligence approaches: firefly algorithm, cuckoo search algorithm, and bat algorithm. We briefly discuss how these methods can be applied to tackle this issue. Some examples of application recently reported in the literature are briefly described. Finally, we envision some future trends and promising lines of research in the field.

A. Iglesias (✉) · A. Gálvez
Department of Applied Mathematics and Computational Sciences,
University of Cantabria, Avda. de Los Castros, s/n, E-39005 Santander, Spain
e-mail: iglesias@unican.es

A. Gálvez
e-mail: galveza@unican.es

A. Iglesias
Department of Information Science, Faculty of Sciences, Narashino Campus,
Toho University, 2-2-1 Miyama, Funabashi 274-8510, Japan

© Springer International Publishing Switzerland 2016
X.-S. Yang (ed.), *Nature-Inspired Computation in Engineering*,
Studies in Computational Intelligence 637, DOI 10.1007/978-3-319-30235-5_8

151

1   Introduction

1.1   *Reverse Engineering*

As players of our exciting digital era, we are all immersed in the worldwide phenomenon of globalization. One of the most noticeable effects of this process is the constant demand of more competitive products with better quality and lower prices. As a result, manufacturers are constantly pressured to design products with new innovative features in order to attract customers attention and increase their sales. One of the strategies to tackle this issue is the mass customization of products. More and more often, manufacturers offer the customers the possibility to customize their products to meet customer's personal preferences and needs. Obviously, this requires a continuous modification of the shape, size, and other geometric attributes of the product. As a consequence, product design is becoming more relevant in the manufacturing process.

Reverse engineering is a key technology in this process. While conventional engineering transforms concepts and models into real objects, reverse engineering operates in the opposite direction. What reverse engineering aims to do is to obtain a digital model of an already existing physical object or component. There are many reasons for this purpose: to analyze how a new machine in the market is built, how a new device or component really works, or to determine whether or not a new product infringes a patent or intellectual property rights. In this context, reverse engineering is a common practice nowadays in consumer products and electronic components. It is also widely applied in many manufacturing industries such as automotive, aerospace, and shipbuilding [1, 2]. A classical approach for the initial conceptual design in these fields is to build prototypes of different workpieces of a model (e.g., car body, airplane fuselage, ship hull) on clay, rubber, foam, wood, metal, or other materials to explore new ideas about shape and size, and allow the designers to interact with the physical model. These prototypes are then digitized by using scanning technologies such as 3D laser scanners, touch scanners, coordinate measuring machines, structured light digitizers, industrial computer tomography scanning, and many others.

The most common outcome of this scanning process is a large collection of measured points, usually represented as a *point cloud* typically comprised of hundreds of thousands, and even millions of data points. This point cloud lacks topological and geometrical information beyond the data points, and must therefore be further processed. The classical approach is to create a model from data by using either a polygonal (usually triangular) mesh, a constructive solid model, or a set of mathematical curves and surfaces. In this chapter we will focus on the latter case, usually referred to as *curve and surface reconstruction*. Classical mathematical functions used for this task are the so-called parametric free-form shapes such as Bézier, B-splines and NURBS curves and surfaces [3]. They consist of a linear combination of a set of basis functions (usually called *blending functions*) where the coefficients (usually called the *poles* or *control points*) can readily be used to control the shape of the fitting function, an extremely valuable feature for interactive design. In general, these blending functions can be classified into two groups: global-support functions and local-support functions. In the former case, the support (the subset of the function domain where the function does not vanish) of the blending functions is the whole domain of the problem, while in the latter case different blending functions can have different supports, which are usually a strict subset of the whole domain. As a consequence, global-support fitting functions exhibit *global control*: any modification of the shape of the curve/surface by moving any pole is automatically propagated throughout the whole curve/surface, something that does not happen in general for local-support fitting functions. Because of limitations of space, in this chapter we will only discuss the case of global-support fitting functions, whereas the local-support case will be the subject of a future paper.

In real-world problems, data points are usually affected by measurement noise, irregular sampling, and other artifacts [1, 2, 4]. Consequently, a good fitting of data should be generally based on approximation schemes rather than on interpolation. This is also the approach taken in this chapter. In this case, the approximating curve or surface is not required to pass through all input data points, but just near to them, according to a given distance criterion.

1.2 Aims and Structure of this Chapter

Although interpolation of all data points is not required, the data fitting problem is still very challenging, as it actually leads to a difficult nonlinear continuous optimization problem. In fact, although several methods have been reported in the literature to tackle this issue (see Sect. 2.4 for details), the problem is still unsolved at large extent. In this chapter, we will discuss some of the most recent and exciting advances in the field as well as some future trends.

The structure of this chapter is as follows: in Sect. 2 we provide the mathematical background required to understand in detail the data fitting problem for curves and surfaces. The section also reports some classical methodologies to solve the resulting optimization problem. In Sect. 3 we focus on nature-inspired metaheuristics and explain the main reasons for applying them to our problem. This discussion is illustrated in Sect. 4 through three exciting nature-inspired optimization techniques recently reported in the literature, namely, the firefly algorithm, the cuckoo search algorithm, and the bat algorithm. The application of these techniques to curve and surface approximation from data points is briefly discussed in Sect. 5. Section 6 provides some hints about future trends in the field. Finally, Sect. 7 summarizes the main conclusions of this chapter.

2   Description of the Problem

This section provides the mathematical formalism of our data fitting problem by using
parametric free-form functions. Firstly, some basic mathematical concepts and definitions are given. Then, the problems of data fitting with global-support curves and
surfaces are discussed, respectively. Finally, we report some previous methodologies
to solve the resulting systems of equations of the least-squares optimization problem.
For further details, we refer the interested reader to the nice introductory book [3].

2.1   Mathematical Concepts and Definitions

A *parametric free-form polynomial curve* $\mathbf{D}(t)$ in $\mathbb{R}^d$ can be mathematically
represented as a finite linear combination of the so-called blending functions
$\{F_j(t)\}_{j=0,\ldots,n}$ as

$$\mathbf{D}(t) = \sum_{j=0}^{n} \mathbf{d}_j F_j(t) \tag{1}$$

where $\mathbf{d}_j$ are vector coefficients in $\mathbb{R}^d$ (usually referred to as the *poles* or *control
points*), and $\{F_j(t)\}_{j=0,\ldots,n}$ is a family of blending functions of parameter t assumed
to be linearly independent and form a basis of the vector space of functions of degree
n on a finite interval $[a, b] \subset \mathbb{R}$. Without loss of generality, we can also assume that
$[a, b]$ is the unit interval $[0, 1]$. Note that in this chapter vectors are denoted in bold.

In this chapter we consider the case in which all blending functions $F_j(t)$ in Eq. (1)
have their support on the whole domain $[a, b]$. In practical terms, this means that the
blending functions provide a global control of the shape of the approximating curve
(these functions are usually referred to as *global-support functions*), as opposed to
the alternative case of local control given by the piecewise representation that is
characteristic of popular curves such as B-splines and NURBS.

A classical example of global-support functions is the canonical polynomial basis:
$p_j(t) = t^j$. Other examples include the Hermite polynomial basis, the trigonometric
basis, the hyperbolic basis, the radial basis, the polyharmonic basis, and many others.
Among them, the most popular ones are the Bernstein polynomials, which are widely
applied in computer graphics, CAD/CAM, and many other industrial fields. In this
case, Eq. (1) corresponds to the *parametric free-form Bézier curve* $\mathbf{C}^p(t)$ *of degree
n*, defined as

$$\mathbf{C}^p(t) = \sum_{j=0}^{n} \mathbf{P}_j B_j^n(t) \tag{2}$$

where $\mathbf{P}_j$ are the control points, and $B_j^n(t)$ are the *Bernstein polynomials of index j and degree n*, given by

$$B_j^n(t) = \binom{n}{j} t^j (1-t)^{n-j} \tag{3}$$

where, by convention, $0! = 1$. The superscript p in Eq. (2) is used to indicate that the corresponding curve is polynomial. An interesting extension of the polynomial Bézier curve is given by the rational case, where the control points are assigned different scalar coefficients called *weights*. In this case, we have a *free-form rational Bézier curve* $\mathbf{C}^r(t)$ *of degree n*, defined as

$$\mathbf{C}^r(t) = \frac{\displaystyle\sum_{j=0}^{n} w_j \mathbf{P}_j B_j^n(t)}{\displaystyle\sum_{j=0}^{n} w_j B_j^n(t)} \tag{4}$$

where w_j denote scalar positive values (the weights) associated with the corresponding control points $\mathbf{P}_j$. Note that if all scalar weights w_j are equal, Eq. (4) becomes Eq. (2) i.e., the rational case transforms into a polynomial one.

A *parametric free-form polynomial Bézier surface of degree (m,n)* is defined as

$$\mathbf{S}(u,v) = \sum_{i=0}^{m} \sum_{j=0}^{n} \mathbf{P}_{i,j} B_i^m(u) B_j^n(v) \tag{5}$$

where $\mathbf{P}_{i,j}$ are the control points arranged in a rectangular grid, (u, v) are the surface parameters, defined on the unit square $[0, 1] \times [0, 1]$, and $B_i^m(u)$ and $B_j^n(v)$ are the Bernstein polynomials of index i and degree m and index j and degree n, respectively.

2.2 Curve Data Fitting Problem

Let us suppose now that we are given a set of data points $\{\mathbf{Q}_i\}_{i=1,\dots,\nu}$ in $\mathbb{R}^d$ (usually $d = 2$ or $d = 3$). Our goal is to obtain the parametric free-form Bézier curve $\mathbf{C}(t)$ (either polynomial $\mathbf{C}^p(t)$ or rational $\mathbf{C}^r(t)$) fitting the data points. This means that we have to compute either the data parameters t_i or the data parameters and weights so that

$$\mathbf{C}^p(t_i) = \sum_{j=0}^{n} \mathbf{P}_j B_j^n(t_i) = \mathbf{Q}_i \qquad i = 1, \dots, \nu \tag{6}$$

or

$$C^r(t_i) = \frac{\sum_{j=0}^{n} w_j \mathbf{P}_j B_j^n(t_i)}{\sum_{j=0}^{n} w_j B_j^n(t_i)} = \mathbf{Q}_i \qquad i = 1, \ldots, \nu \tag{7}$$

respectively, by solving the corresponding systems of equations. In many real-world applications, the number of data points ν is very large (of order 10^5 to 10^6 and even larger). For practical reasons, we expect to find a fitting curve with many fewer parameters, i.e. $n \ll \nu$, meaning that the system (6) (or (7), respectively) is over-determined. In this case, no analytical solution can be obtained. Instead, the goal is to compute the fitting function that minimizes the least-squares (LSQ) error, E, defined as the sum of squares of the residuals. This LSQ error is given by either

$$E^p = \sum_{i=1}^{\nu} \left(\mathbf{Q}_i - \sum_{j=0}^{n} \mathbf{P}_j B_j^n(t_i) \right)^2 \tag{8}$$

or

$$E^r = \sum_{i=1}^{\nu} \left(\mathbf{Q}_i - \frac{\sum_{j=0}^{n} w_j \mathbf{P}_j B_j^n(t_i)}{\sum_{j=0}^{n} w_j B_j^n(t_i)} \right)^2 \tag{9}$$

for the polynomial and the rational cases, respectively, where we need a parameter value t_i to be associated with each data point $\mathbf{Q}_i$, $i = 1, \ldots, \nu$. In the first case, considering the column vectors $\mathbf{B}_j = (B_j^n(t_1), \ldots, B_j^n(t_\nu))^T$, $j = 0, \ldots, n$, where $(.)^T$ means transposition, and $\bar{\mathbf{Q}} = (\mathbf{Q}_1, \ldots, \mathbf{Q}_\nu)$, Eq. (8) becomes the following system of equations (called the *normal equation*):

$$\begin{pmatrix} \mathbf{B}_0^T.\mathbf{B}_0 & \ldots & \mathbf{B}_n^T.\mathbf{B}_0 \\ \vdots & \vdots & \vdots \\ \mathbf{B}_0^T.\mathbf{B}_n & \ldots & \mathbf{B}_n^T.\mathbf{B}_n \end{pmatrix} \begin{pmatrix} \mathbf{P}_0 \\ \vdots \\ \mathbf{P}_n \end{pmatrix} = \begin{pmatrix} \bar{\mathbf{Q}}.\mathbf{B}_0 \\ \vdots \\ \bar{\mathbf{Q}}.\mathbf{B}_n \end{pmatrix} \tag{10}$$

which can be compacted as:

$$\mathbf{M}\mathbf{P} = \mathbf{Q} \tag{11}$$

with $\mathbf{M} = \left[\sum_{j=1}^{\nu} B_l^n(t_j) B_i^n(t_j) \right]$, $\mathbf{Q} = \left[\sum_{j=1}^{\nu} \mathbf{Q}_j B_l^n(t_j) \right]$ for $i, l = 0, \ldots, n$.

A similar discussion applies for the rational case by taking the rational blending functions:

$$\mathbf{R}_k(t) = \frac{w_k B_k^n(t)}{\sum_{j=0}^{n} w_j B_j^n(t)} \qquad k = 0, \ldots, n \tag{12}$$

In this case, Eq. (9) can also be expressed as Eqs. (10)–(11) (the mathematical details are left to the reader).

2.3 Surface Data Fitting Problem

In the case of surfaces, let us suppose now that we are given a set of (organized) data points $\{\mathbf{Q}_{k,l}\}_{k=1,\ldots,\mu; l=1,\ldots,\nu}$ in $\mathbb{R}^3$. In this case, the goal is to obtain the free-form Bézier surface $\mathbf{S}(u, v)$ that fits the data points better in the discrete least-squares sense. Once again, this is done by minimizing the least-squares error, E^S, defined as the sum of squares of the residuals:

$$E^S = \sum_{k=1}^{\mu} \sum_{l=1}^{\nu} \left(\mathbf{Q}_{k,l} - \sum_{i=0}^{m} \sum_{j=0}^{n} \mathbf{P}_{i,j} B_i^m(u_k) B_j^n(v_l) \right)^2 \tag{13}$$

In the case of irregularly sampled data points $\{\mathbf{Q}_r\}_{r=1,\ldots,\kappa}$, our method will work in a similar way by simply replacing the previous expression (13) by:

$$E^S = \sum_{r=1}^{\kappa} \left(\mathbf{Q}_r - \sum_{i=0}^{m} \sum_{j=0}^{n} \mathbf{P}_{i,j} B_i^m(u_r) B_j^n(v_r) \right)^2 \tag{14}$$

The least-squares minimization of either (13) or (14) leads to the system of equations:

$$\langle \mathbf{Q} \rangle = \langle \mathbf{P} \rangle \cdot \varXi \tag{15}$$

where $\langle \mathbf{Q} \rangle$ corresponds to the vectorization of the matrix (the column vector obtained by stacking the columns of the matrix on top of one another) of data points $\{\mathbf{Q}_{k,l}\}_{k=1,\ldots,\mu; l=1,\ldots,\nu}$ (alternatively, $\{\mathbf{Q}_r\}_{r=1,\ldots,R}$), $\langle \mathbf{P} \rangle$ corresponds to the vectorization of the set of control points $\{\mathbf{P}_{i,j}\}_{i=0,\ldots,m; j=0,\ldots,n}$, and $\varXi$ is a matrix given by: $\varXi_{i,j} = \mathbf{B}^n(v_j) \odot B_0^m(\mathbf{u})$, with $\mathbf{B}^h(\omega_k) = (B_0^h(\omega_k), \ldots, B_H^h(\omega_k))$, $B_k^h(\Theta) = (B_k^h(\theta_1), \ldots, B_k^h(\theta_K))$, for any $\Theta = (\theta_1, \ldots, \theta_K)$, "." represents the matrix product, and $\odot$ represents the tensor product of vectors.

2.4 Solving the System of Equations

Note that, since the Bernstein polynomials are nonlinear in their variables, the minimization of the least-squares error is a nonlinear continuous optimization problem. As discussed above, we are also dealing with a high-dimensional problem. In many cases, it is also a multimodal problem, since there might be more than one solution leading to the optimal value for the fitness function. Therefore, solving the systems of equations (6) or (7) for curves or (13) or (14) for surfaces leads to a very difficult multimodal, multivariate, continuous, nonlinear optimization problem. Furthermore, we are interested to solve the general problem, meaning that all parameters of the problem must be computed at full extent.

This general problem has been addressed for many years. First approaches in the 60 s and 70 s were mostly based on numerical procedures [3, 5, 6]. However, it was found that traditional mathematical optimization techniques are only well suited for some particular cases but fail to solve the general problem. As a consequence, the scientific community in the field turned its attention to other alternative approaches to this problem. Some examples are the error bounds [7], curvature-based squared distance minimization [8], or dominant points [9] techniques. In general, they perform well but are restricted to particular cases, as they require conditions (high differentiability, closed curves, noiseless data) which are not very common in real-world applications.

An interesting approach consists of applying artificial intelligence techniques to this problem [10–12], such as standard neural networks [11], or Kohonen's SOM (Self-Organizing Maps) nets [12–14]. In some cases, this neural approach is combined with partial differential equations [10] or other approaches [15, 16]. However, the neural networks are severely limited in many ways. On one hand, it is difficult to describe the functional structure of the problem by using only scalar values. This problem can be overcome by using functional networks, a powerful generalization of the neural networks where the scalar weights are replaced by functions [17–19]. On the other hand, they require to specify an initial topology for the network, a difficult issue when little or no information about the problem is available.

A promising recent line of research is based on the application of nature-inspired metaheuristic techniques, which have been intensively applied to solve difficult optimization problems that cannot be tackled through traditional optimization algorithms. Genetic algorithms have been applied to data fitting in both the discrete version [20] and the continuous version [21, 22]. Other metaheuristic approaches applied to this issue include the popular particle swarm optimization technique [23–25], simulated annealing [26], support vector machines [27], artificial immune systems [28–30], estimation of distribution algorithms [31], taboo search [32], genetic programming [33], memetic methods [34], and hybrid techniques [20, 35, 36]. In this chapter, we will explore the potential of some very recent nature-inspired metaheuristic techniques to solve the data fitting problem with free-form curves and surfaces. They will be described in Sect. 4. Before, a gentle overview about nature-inspired metaheuristics is given in next section.

3 Nature-Inspired Metaheuristics

Among the wealth of optimization methods developed during the last decades, one of the most promising set of techniques is given by the nature-inspired metaheuristics. By *metaheuristics* we refer to a high-level procedure or strategy designed to assist a low-level search strategy or heuristic to find a good solution for an optimization problem. Of course, this notion does not denote a single method, or even a single family of methods. Instead, it encompasses many kinds of techniques, which are very different in nature but still share a common feature: they usually make very few (if any) assumptions about the problem to be solved. As a result, they are very general and versatile and can be applied to many different optimization problems with only minor modifications (e.g., parameter tuning). A major limitation of metaheuristics, however, is that no global optimum solution is fully guaranteed.

In many cases, metaheuristics are inspired by the behavior of social living organisms in nature, a sub-field typically referred to as *swarm intelligence* [37–39]. In this case, a given initial population is modified along successive iterations (called generations) according to some rules or evolution equations which are idealizations or simplifications of those dictating the dynamical evolution of the real swarms in nature. Typical examples of these nature-inspired swarm intelligence methods are ant colony optimization (ACO), genetic algorithms (GA), particle swarm optimization (PSO), artificial bee colony (ABC), and many others [40–42].

In this chapter, we are particularly interested in nature-inspired swarm intelligence metaheuristics, as they are very well suited to our problem. There are four main reasons that explain why it is advisable to apply them to solve the curve/surface reconstruction problem:

1. *they can be used even when we have very little information about the problem.* Many papers reported in the literature apply alternative methods to some academic examples in this field. However, such examples have some kind of topological or geometric structure, which is advantageously used in the method to solve the problem. This is not the case, however, in real-world reverse engineering applications, where typically little or no information about the problem is known beyond the data points.
2. *their objective function does not need to be continuous or differentiable.* In general, metaheuristics have shown to be able to cope with optimization problems with underlying functions which are non-differentiable (even non-continuous in some instances).
3. *they can be successfully applied to multimodal and multivariate nonlinear optimization problems.* Nature-inspired swarm intelligence metaheuristics are very well suited to find optimal or near-to-optimal solutions to nonlinear optimization problems in high-dimensional search spaces. The curve and surface data fitting problems described in Sects. 2.2 and 2.3 are two of such problems. Furthermore, owing to the (potential) existence of many local optima of the least-squares objective function, very often they are also multimodal problems. These techniques are also well suited for multimodal problems.

4. *they can deal with noisy data*. This is a very important issue in many real-world applications. For instance, laser scanner systems yield an enormous amount of irregular data points. The same applies to other data capturing methods as well. As a consequence, reconstruction methods must be robust against this measurement noise, irregular sampling, and other artifacts.

4   Some Recent Nature-Inspired Swarm Intelligence Metaheuristics

During the last few years, we have witnessed the staggering appearance and development of many new nature-inspired swarm intelligence metaheuristic techniques. Although not all of them have actually shown a real usefulness and the ability to improve previous results for some optimization problems, we can certainly say that currently this is a very active field of research in computational intelligence. Among the huge amount of new methods reported in the literature, in this section we will focus on three interesting methods that, albeit introduced very recently, have already proved their power to solve complex optimization problems pretty efficiently. Furthermore, they have already been applied to the data fitting problem discussed in this chapter. In next subsections, we will provide the reader with a succinct description about the basics of the firefly algorithm, cuckoo search algorithm, and bat algorithm, respectively. The interested reader is referred to the nice books in [43, 44] for an in-depth description of these methods, their variants, implementation, and applications.

4.1   The Firefly Algorithm

The firefly algorithm is a nature-inspired metaheuristic algorithm introduced in 2008 by Xin-She Yang to solve optimization problems [43, 45, 46]. The algorithm is based on the social flashing behavior of fireflies in nature. The key ingredients of the method are the variation of light intensity and formulation of attractiveness. In general, the attractiveness of an individual is assumed to be proportional to their brightness, which in turn is associated with the encoded objective function. In the firefly algorithm, there are three particular idealized rules, which are based on some of the major flashing characteristics of real fireflies [45]. They are:

1. All fireflies are unisex, so that one firefly will be attracted to other fireflies regardless of their sex;
2. The degree of attractiveness of a firefly is proportional to its brightness, which decreases as the distance from the other firefly increases due to the fact that the air absorbs light. For any two flashing fireflies, the less brighter one will move towards the brighter one. If there is not a brighter or more attractive firefly than a particular one, it will then move randomly;

3. The brightness or light intensity of a firefly is determined by the value of the objective function of a given problem. For instance, for maximization problems, the light intensity can simply be proportional to the value of the objective function.

The distance between any two fireflies i and j, at positions $\mathbf{X}_i$ and $\mathbf{X}_j$, respectively, can be defined as a Cartesian or Euclidean distance as follows:

$$r_{ij} = ||\mathbf{X}_i - \mathbf{X}_j|| = \sqrt{\sum_{k=1}^{D}(x_{i,k} - x_{j,k})^2} \tag{16}$$

where $x_{i,k}$ is the k-th component of the spatial coordinate $\mathbf{X}_i$ of the i-th firefly and D is the number of dimensions.

In the firefly algorithm, as attractiveness function of a firefly j one should select any monotonically decreasing function of the distance to the chosen firefly, e.g., the exponential function:

$$\beta = \beta_0 e^{-\gamma r_{ij}^{\mu}} \quad (\mu \geq 1) \tag{17}$$

where r_{ij} is the distance defined as in Eq. (16), β_0 is the initial attractiveness at $r = 0$, and γ is an absorption coefficient at the source which controls the decrease of the light intensity.

The movement of a firefly i which is attracted by a more attractive (i.e., brighter) firefly j is governed by the following evolution equation:

$$\mathbf{X}_i = \mathbf{X}_i + \beta_0 e^{-\gamma r_{ij}^{\mu}} (\mathbf{X}_j - \mathbf{X}_i) + \alpha \left(\sigma - \frac{1}{2}\right) \tag{18}$$

where the first term on the right-hand side is the current position of the firefly, the second term is used for considering the attractiveness of the firefly to light intensity seen by adjacent fireflies, and the third term is used for the random movement of a firefly in case there are not any brighter ones. The coefficient α is a randomization parameter determined by the problem of interest, while σ is a random number generator uniformly distributed in the space [0, 1]. The corresponding pseudo-code of the firefly algorithm is shown in Table 1.

4.2 The Cuckoo Search Method

Cuckoo search (CS) is a nature-inspired population-based metaheuristic algorithm originally proposed by Yang and Deb in 2009 to solve optimization problems [47]. The algorithm is inspired by the obligate interspecific brood-parasitism of some cuckoo species that lay their eggs in the nests of host birds of other species with the aim of escaping from the parental investment in raising their offspring. This strategy is also useful to minimize the risk of egg loss to other species, as the cuckoos can distributed their eggs amongst a number of different nests. Of course, sometimes it happens that the host birds discover the alien eggs in their nests. In such cases,

Table 1 Firefly algorithm pseudocode

Algorithm: Firefly Algorithm
begin
Objective function $f(\mathbf{x})$, $\mathbf{x} = (x_1, \ldots, x_D)^T$
Generate initial population of n fireflies $\mathbf{x}_i$ $(i = 1, 2, \ldots, n)$
Formulate light intensity I associated with $f(\mathbf{x})$
Define absorption coefficient γ
while $(t < MaxGeneration)$ or (stop criterion)
for $i = 1$ to n
for $j = 1$ to n
if $I(i) > I(j)$
then move firefly i towards j
end if
Vary attractiveness with distance r via $e^{-\gamma r}$
Evaluate new solutions and update light intensity
end for
end for
Rank fireflies and find the current best
end while
Post-processing the results and visualization
end

the host bird can take different responsive actions varying from throwing such eggs away to simply leaving the nest and build a new one elsewhere. However, the brood parasites have at their turn developed sophisticated strategies (such as shorter egg incubation periods, rapid nestling growth, egg coloration or pattern mimicking their hosts, and many others) to ensure that the host birds will care for the nestlings of their parasites.

This interesting and surprising breeding behavioral pattern is the metaphor of the cuckoo search metaheuristic approach for solving optimization problems. In the cuckoo search algorithm, the eggs in the nest are interpreted as a pool of candidate solutions of an optimization problem while the cuckoo egg represents a new coming solution. The ultimate goal of the method is to use these new (and potentially better) solutions associated with the parasitic cuckoo eggs to replace the current solution associated with the eggs in the nest. This replacement, carried out iteratively, will eventually lead to a very good solution of the problem.

In addition to this representation scheme, the CS algorithm is also based on three idealized rules [47, 48]:

1. Each cuckoo lays one egg at a time, and dumps it in a randomly chosen nest;
2. The best nests with high quality of eggs (solutions) will be carried over to the next generations;
3. The number of available host nests is fixed, and a host can discover an alien egg with a probability $p_a \in [0, 1]$. In this case, the host bird can either throw the egg away or abandon the nest so as to build a completely new nest in a new location.

Table 2 Cuckoo search algorithm via Lévy flights as originally proposed by Yang and Deb in [47, 48]

Algorithm: Cuckoo Search via Lévy Flights
begin
Objective function $f(\mathbf{x})$, $\mathbf{x} = (x_1, \ldots, x_D)^T$
Generate initial population of n host nests $\mathbf{x}_i$ $(i = 1, 2, \ldots, n)$
while $(t < MaxGeneration)$ or (stop criterion)
Get a cuckoo (say, i) randomly by Lévy flights
Evaluate its fitness F_i
Choose a nest among n (say, j) randomly
if $(F_i > F_j)$
Replace j by the new solution
end
A fraction (p_a) of worse nests are abandoned and new ones
are built via Lévy flights
Keep the best solutions (or nests with quality solutions)
Rank the solutions and find the current best
end while
Postprocess results and visualization
end

For simplicity, the third assumption can be approximated by a fraction p_a of the n nests being replaced by new nests (with new random solutions at new locations). For a maximization problem, the quality or fitness of a solution can simply be proportional to the objective function. However, other (more sophisticated) expressions for the fitness function can also be defined.

Based on these three rules, the basic steps of the CS algorithm can be summarized as shown in the pseudo-code reported in Table 2. Basically, the CS algorithm starts with an initial population of n host nests and it is performed iteratively. In the original proposal, the initial values of the jth component of the ith nest are determined by the expression $x_i^j(0) = rand \cdot (up_i^j - low_i^j) + low_i^j$, where up_i^j and low_i^j represent the upper and lower bounds of that jth component, respectively, and $rand$ represents a standard uniform random number on the open interval $(0, 1)$. Note that this choice ensures that the initial values of the variables are within the search space domain. These boundary conditions are also controlled in each iteration step.

For each iteration g, a cuckoo egg i is selected randomly and new solutions $\mathbf{x}_i(g + 1)$ are generated by using the Lévy flight, a kind of random walk in which the steps are defined in terms of the step-lengths, which have a certain probability distribution, with the directions of the steps being isotropic and random. According to the original creators of the method, the strategy of using Lévy flights is preferred over other simple random walks because it leads to better overall performance of the CS. The general equation for the Lévy flight is given by:

$$\mathbf{x}_i(g + 1) = \mathbf{x}_i(g) + \alpha \oplus levy(\lambda) \tag{19}$$

where g indicates the number of the current generation, and $\alpha > 0$ indicates the step size, which should be related to the scale of the particular problem under study. The symbol $\oplus$ is used in Eq. (19) to indicate the entry-wise multiplication. Note that Eq. (19) is essentially a Markov chain, since next location at generation $g + 1$ only depends on the current location at generation g and a transition probability, given by the first and second terms of Eq. (19), respectively. This transition probability is modulated by the Lévy distribution as:

$$levy(\lambda) \sim g^{-\lambda}, \quad (1 < \lambda \leq 3) \tag{20}$$

which has an infinite variance with an infinite mean. Here the steps essentially form a random walk process with a power-law step-length distribution with a heavy tail. From the computational standpoint, the generation of random numbers with Lévy flights is comprised of two steps: firstly, a random direction according to a uniform distribution is chosen; then, the generation of steps following the chosen Lévy distribution is carried out. The authors suggested to use the so-called Mantegna's algorithm for symmetric distributions, where "symmetric" means that both positive and negative steps are considered (see [43] for details). Their approach computes the factor:

$$\hat{\phi} = \left(\frac{\Gamma(1 + \hat{\beta}) \cdot \sin\left(\frac{\pi \cdot \hat{\beta}}{2}\right)}{\Gamma\left(\left(\frac{1+\hat{\beta}}{2}\right) \cdot \hat{\beta} \cdot 2^{\frac{\hat{\beta}-1}{2}}\right)} \right)^{\frac{1}{\hat{\beta}}} \tag{21}$$

where Γ denotes the Gamma function and $\hat{\beta} = \dfrac{3}{2}$ in the original implementation by Yang and Deb [48]. This factor is used in Mantegna's algorithm to compute the step length ς as:

$$\varsigma = \frac{u}{|v|^{\frac{1}{\beta}}} \tag{22}$$

where u and v follow the normal distribution of zero mean and deviation σ_u^2 and σ_v^2, respectively, where σ_u obeys the Lévy distribution given by Eq. (21) and $\sigma_v = 1$. Then, the stepsize η is computed as:

$$\eta = 0.01 \, \varsigma \, (\mathbf{x} - \mathbf{x}_{best}) \tag{23}$$

where ς is computed according to Eq. (22). Finally, $\mathbf{x}$ is modified as: $\mathbf{x} \leftarrow \mathbf{x} + \eta \cdot \Upsilon$ where Υ is a random vector of the dimension of the solution $\mathbf{x}$ and that follows the normal distribution $N(0, 1)$.

The CS method then evaluates the fitness of the new solution and compares it with the current one. In case the new solution brings better fitness, it replaces the current one. On the other hand, a fraction of the worse nests (according to the fitness) are abandoned and replaced by new solutions so as to increase the exploration of the search space looking for more promising solutions. The rate of replacement is given by the probability p_a, a parameter of the model that has to be tuned for better performance. Moreover, for each iteration step, all current solutions are ranked according to their fitness and the best solution reached so far is stored as the vector $\mathbf{x}_{best}$ (used, for instance, in Eq. (23)).

This algorithm is applied in an iterative fashion until a stopping criterion is met. Common terminating criteria are that a solution is found that satisfies a lower threshold value, or that a fixed number of generations has been reached, or that successive iterations no longer produce better results.

4.3   The Bat Algorithm

The bat algorithm is a bio-inspired population-based meta-heuristic algorithm originally proposed by Xin-She Yang in 2010 to solve difficult optimization problems [49, 50]. The algorithm is based on the echolocation behavior of microbats, which use a type of sonar called echolocation, with varying pulse rates of emission and loudness, to detect prey, avoid obstacles, and locate their roosting crevices in the dark. The interested reader is referred to the general paper in [51] for a comprehensive, updated review of the bat algorithm and all its variants and applications. See also [52] for its application to multi-objective optimization. The algorithm can be summarized as follows:

1. Bats use echolocation to sense distance and distinguish between food, prey, and background barriers.
2. Each virtual bat flies randomly with a velocity $\mathbf{v}_i$ at position (solution) $\mathbf{x}_i$ with a fixed frequency f_{min}, varying wavelength λ and loudness A_0 to search for prey. As it searches and finds its prey, it changes wavelength (or frequency) of their emitted pulses and adjust the rate of pulse emission r, depending on the proximity of the target.
3. It is assumed that the loudness will vary from an (initially large and positive) value A_0 to a minimum constant value A_{min}.

In order to apply the bat algorithm for optimization problems more efficiently, we assume that the frequency f evolves on a bounded interval $[f_{min}, f_{max}]$, meaning that the wavelength λ is also bounded, because $\lambda.f$ is constant. For practical reasons, it is also convenient that the largest wavelength is comparable to the size of the search space. For simplicity, we can assume that $f_{min} = 0$, so $f \in [0, f_{max}]$. The rate of pulse can simply be in the range $r \in [0, 1]$, where 0 means no pulses at all, and 1 means the maximum rate of pulse emission.

The basic pseudo-code of the bat algorithm is shown in Algorithm 1. Basically, the algorithm considers an initial population of $\mathscr{P}$ individuals (bats). Each bat, representing a potential solution of the optimization problem, has a location $\mathbf{x}_i$ and velocity $\mathbf{v}_i$. The algorithm initializes these variables with random values within the search space. Then, the pulse frequency, pulse rate, and loudness are computed for each individual bat (lines 2–6). Then, the swarm evolves in a discrete way over

---

Algorithm 1 Bat Algorithm

Require: (Initial Parameters)
    Population size: $\mathscr{P}$
    Maximum number of generations: $\mathscr{G}_{\max}$
    Loudness: $\mathscr{A}$
    Pulse rate: r
    Maximum frequency: $f_{\max}$
    Dimension of the problem: d
    Objective function: $\phi(\mathbf{x})$, with $\mathbf{x} = (x_1, \ldots, x_d)^T$
    Random vectors: $\Theta = (\theta_1, \ldots, \theta_{\mathscr{P}})$, $\Psi = (\psi_1, \ldots, \psi_{\mathscr{P}})$ with $\theta_k, \psi_k \in U(0, 1)$
1: $g \leftarrow 0$ //g: generation index
2: **for** $i = 1$ **to** $\mathscr{P}$ **do**
3:    Initialize the location and velocity $\mathbf{x}_i$ and $\mathbf{v}_i$ //Initialization phase
4:    Define pulse frequency f_i at $\mathbf{x}_i$
5:    Initialize pulse rates r_i and loudness $\mathscr{A}_i$
6: **end for**
7: **while** $g \leq \mathscr{G}_{\max}$ **do**
8:    **for** $i = 1$ **to** $\mathscr{P}$ **do**
9:        Generate new solutions by adjusting frequency,
10:       and updating locations and velocities   //eqns. (24)-(26)
11:      **if** $\theta_i > r_i$ **then**
12:         $\mathbf{s}^{best} \leftarrow \mathbf{s}^g$ //select the current best global solution
13:         $\mathbf{ls}^{best} \leftarrow \mathbf{ls}^g$ //generate a local solution around $\mathbf{s}^{best}$
14:      **end if**
15:      Generate a new solution by local random walk   //eqn. (27)
16:      **if** $\psi_i \langle \mathscr{A}_i$ and $\phi(\mathbf{x}_i) \langle \phi(\mathbf{x}^*)$ **then**
17:        Accept new solutions
18:        Increase r_i and decrease $\mathscr{A}_i$ //eqns. (28)-(29)
19:      **end if**
20:    **end for**
21:    $g \leftarrow g + 1$
22: **end while**
23: Rank the bats and find current best $\mathbf{x}^*$
24: **return** $\mathbf{x}^*$

---

generations (line 7), like time instances (line 21) until the maximum number of generations, $\mathscr{G}_{\max}$, is reached (line 22). For each generation g and each bat (line 8), new frequency, location and velocity are computed (lines 9–10) according to the following evolution equations:

$$f_i^g = f_{\min}^g + \beta(f_{\max}^g - f_{\min}^g) \tag{24}$$

$$\mathbf{v}_i^g = \mathbf{v}_i^{g-1} + [\mathbf{x}_i^{g-1} - \mathbf{x}^*] f_i^g \tag{25}$$

$$\mathbf{x}_i^g = \mathbf{x}_i^{g-1} + \mathbf{v}_i^g \tag{26}$$

where $\beta \in [0, 1]$ follows the random uniform distribution, and $\mathbf{x}^*$ represents the current global best location (solution), which is obtained through evaluation of the objective function at all bats and ranking of their fitness values. The superscript $(.)^g$ is used to denote the current generation g.

The current global best solution and a local solution around it are probabilistically selected according to a given criterion (lines 11–14). Then, search is intensified by a local random walk (line 15). For this local search, once a solution is selected among the current best solutions, it is perturbed locally through a random walk of the form:

$$\mathbf{x}_{new} = \mathbf{x}_{old} + \varepsilon \mathscr{A}^g \tag{27}$$

where ε is a random number with uniform distribution on the interval $[-1, 1]$ and $\mathscr{A}^g = \langle \mathscr{A}_i^g \rangle$, is the average loudness of all the bats at generation g.

If the new solution achieved is better than the previous best one, it is probabilistically accepted depending on the value of the loudness. In that case, the algorithm increases the pulse rate and decreases the loudness (lines 16–19). This process is repeated for the given number of generations. In general, the loudness decreases once a bat finds its prey (in our analogy, once a new best solution is found), while the rate of pulse emission decreases. For simplicity, the following values are commonly used: $\mathscr{A}_0 = 1$ and $\mathscr{A}_{min} = 0$, assuming that this latter value means that a bat has found the prey and temporarily stop emitting any sound. The evolution rules for loudness and pulse rate are as follows:

$$\mathscr{A}_i^{g+1} = \alpha \mathscr{A}_i^g \tag{28}$$
$$r_i^{g+1} = r_i^0 [1 - exp(-\gamma g)] \tag{29}$$

where α and γ are constants. Note that for any $0 < \alpha < 1$ and any $\gamma > 0$ we have:

$$\mathscr{A}_i^g \to 0, \quad r_i^g \to r_i^0, \quad \text{as } g \to \infty \tag{30}$$

In general, each bat should have different values for loudness and pulse emission rate, which can be computationally achieved by randomization. To this aim, we can take an initial loudness $\mathscr{A}_i^0 \in (0, 2)$ while the initial emission rate r_i^0 can be any value in the interval $[0, 1]$. Loudness and emission rates will be updated only if the new solutions are improved, an indication that the bats are moving towards the optimal solution.

5 Application to the Data Fitting Problem

The three nature-inspired metaheuristics in previous Sects. 4.1–4.3 have been applied to solve different cases of the data fitting problem described in Sect. 2. In this section we describe the important issue of the problem representation and the main conclusions derived from the application of these techniques to our problem. Of course, our discussion is necessarily short due to space limitations. The interested reader is referred to the original sources for further details.

5.1   Representation of the Problem

To solve the data fitting problem, we need a proper representation of the problem.
In the three methods described in previous sections, the population of individuals
(fireflies, eggs, and bats, respectively) represents the pool of candidate solutions to a
given optimization problem. In our particular case, they are therefore associated with
the data parameterization along with the weights (for the rational case only). The
basic idea in this representation is that individuals in the swarm population should
contain at least the variables of the nonlinear subspace of the given problem. Note
that, according to our previous assumptions in Sect. 2, the dimension of this nonlinear
subspace is generally much larger than that of its linear counterpart. Note also that
data parameterization can be performed not only in its continuous version but also
in a discrete fashion. In the latter case, the classical approach is to consider an initial
discretization of the parameter vector and a set of binary strings. Values 1 in each
string are used to indicate that the corresponding discrete values associated with the
positions of the same indices in the solution vector are considered; values 0 mean
that they are simply discarded.

In the case of surfaces, two different approaches are taken depending on whether
the data points are organized or unorganized. In the former case, the candidate solutions are encoded as tensor-product matrices where vectors of candidate solutions
are considered for each isoparametric variable, which are subsequently processed in
a similar way that for the case of curves. Since data points are usually affected by
noise, the tensor-product structure is sometimes damaged or even lost. In this case,
the rectangular topology can still be recovered by using a number of techniques,
the most usual ones so far being the Kohonen's self-organizing maps networks and
classical clustering techniques such as K-means. In the latter case, the candidate
solutions are simply encoded as vectors of strictly increasing real values. Flattening
of the resulting vectors after solving the corresponding system of equations given
by Eq. (15) yields a two-dimensional array whose length is determined by the initial
choice of the surface degree.

5.2   Examples of Application

The problem of data fitting with polynomial Bézier curves has been addressed in
[53–55] by using a firefly algorithm, a cuckoo search algorithm, and a bat algorithm,
respectively. The firefly algorithm was applied to two 2D curves exhibiting challenging features such as several self-intersections and cusps, and a 3D curve (the Viviani
curve). In all cases, the input consists of sets of non-uniformly sampled data points
and affected by a low-intensity random noise perturbation. Furthermore, no information about the parameterization is known in advance. The cuckoo search algorithm
was applied to five examples, including two open 2D curves (Agnessi curve and
Archemedian spiral curve), a self-intersecting curve (hypocycloid), a closed curve

with a cusp (piriform curve) and a closed 3D curve (the Eight Knot curve). In all examples, data points were perturbed by an additive Gaussian white noise of low intensity given by a SNR (signal-to-noise ratio) of 60. The approximation error was of order 10^{-2} to 10^{-3} depending on the example. The bat algorithm was applied to the examples of epitrochoid, piriform curve and Viviani curve, already used in previous papers with other methods. Although the fitting errors with the bat algorithm reported in that paper are much better than in previous references, the results cannot be actually compared because data points in this case are not affected by measurement noise. However, the paper also included a comparative work with the firefly algorithm and the arc-length parameterization for the proposed benchmark and the same (noiseless) conditions. The main conclusion was that the results achieved by the firefly algorithm and the bat algorithm are pretty similar, and both outperform those obtained with the arc-length parameterization by orders of magnitude. As an illustration, Fig. 1 shows two examples of data fitting with polynomial Bézier curves by using the firefly algorithm (left) and the bat algorithm (right). In both pictures, the original data points are represented by green squares and the fitting curve as a blue solid line. The corresponding parameters for these methods and the fitting errors are reported in Table 3.

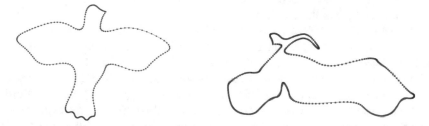

Fig. 1 Two examples of data fitting with polynomial Bézier curves: (*left*) a flying pigeon; (*right*) a motorbike. The method is the firefly algorithm and the bat algorithm, respectively (see Table 3 for further details about these examples)

Table 3 Benchmark shown in this chapter along with the main features of each example

Example	Flying pigeon	Motorbike
# Data points	250	400
Optimization method	Firefly algorithm	Bat algorithm
Parameter values	$\alpha = 0.1$	$\mathscr{A} = 0.5$
	$\beta = 1$	$r = 0.5$
	$\gamma = 0.5$	$f_{max} = 2$
Fitting errors	$x : 5.764451E\text{-}4$	$x : 1.927563E\text{-}3$
	$y : 4.732029E\text{-}4$	$y : 2.153489E\text{-}3$

This polynomial Bézier curve case has been very recently extended to the rational Bézier case for the firefly algorithm and the bat algorithm in [56, 57], respectively. In [56], eight different variants of the firefly algorithm have been applied to a benchmark of four examples: the epicycloid, a bulb light, a sakura flower, and a cat silhouette. The variants are obtained by all possible combinations of three different criteria (memetic, elitism, and rational) being set *on/off*, so the polynomial case is also included in the discussion. The main conclusion is that the variant with value *on* for all criteria outperforms all other alternatives for all instances in the benchmark. Moreover, we determined that the rational criterion is the most important one, with elitism being the less relevant one. Finally, the hybridization of the firefly algorithm with a local search procedure helps to improve the quality of the solutions, but not dramatically. The work in [57] applied the bat algorithm for the rational scheme to three examples: two 2D curves (hypocycloid and spiral) and a 3D curve (helix). The obtained results for this scheme are very good, with the RMSE (root-mean-square error) ranging from 10^{-4} (average) to 10^{-5} (best) for the three examples in the benchmark. This excellent performance is not very surprising, as the rational scheme is significantly more powerful than the polynomial one. The ultimate reason for this superiority is the fact that the rational case provides additional variables, i.e. extra degrees of freedom that can be used for better fitting. However, this desirable feature does not come for free; these additional variables are also to be computed, thus increasing the computational load of the problem.

The data fitting problem for polynomial Bézier surfaces has been addressed in [58, 59] with the firefly algorithm and the bat algorithm, respectively. In the first case, the input consists of sets of irregularly sampled data points affected by measurement noise of low to medium intensity (signal-to-noise ratio of 15, 25, and 10, for the three examples considered). The fitting errors are pretty good, with RMSE of order 10^{-3} in all cases, for examples that are noisy and highly oscillating and exhibit a rich variety of hills and valleys. The second example was particularly challenging because it is a closed surface with a strong singularity at its uppermost part, where many data points concentrate in a very small volume. Most the methods fail in this case, because free-form parametric surfaces typically tend to distribute the control points by following a rectangular topology. Obviously, such a distribution is not adequate for that example. However, the firefly algorithm solves the problem automatically and rearranges the control points by itself to adapt to the underlying structure of data points. This is a remarkable result that shows the ability of the method to capture the real behavior of data points even under unfavorable conditions. The paper in [59] also considers three clouds of data points from which three polynomial Bézier surfaces are reconstructed by using the bat algorithm. RMSE errors in this case are of order 10^{-4} for the best cases. The results of both papers should not be compared, however, because of two main reasons: firstly, they use different benchmarks and, secondly, data points in the second paper are not affected by measurement noise. In our opinion, the determination of a suitable benchmark for comparative purposes is a must, and one of our proposals for future trends in the field, as discussed in next section.

6 Future Trends

In spite of the extensive research work carried out so far, the curve/surface reconstruction problem is still far from being solved in all its generality. For decades, the search of a solution was focused on traditional mathematical techniques, and significant time and effort was devoted to this aim with little success. At that point it became clear that we are actually dealing with a very difficult optimization problem, much more difficult than it was initially guessed. As a result, during the last two decades or so the research community has shifted their attention to other alternative methods, of which the nature-inspired metaheuristics have shown to be the most promising ones. But, despite the good results achieved so far for this kind of methods, there is still a long way to walk. We are just scratching the surface by considering global-support approaches, which are likely not the most efficient ones in solving the problem. A future trend is the extension of this methodology to local-support functions. This demands not only a new formulation of the problem but also more powerful methods, since new parameters and variables are now to be computed in order to solve the general case [60, 61]. In this sense, the development of new, more powerful metaheuristics would be highly beneficial for the field. Also, the combination of existing nature-inspired computational methodologies (such as those reported in this chapter) with other search methods could arguably open the door to improved solutions in terms of fitting errors and computation times. Among these possibilities, the hybridization of swarm intelligence methods with local search methods (the so-called memetic algorithms) might potentially represent a significant step in the way. In our opinion, this tendency will even increase in the future, and we expect a number of memetic algorithms to be applied to this problem.

Another future trend in this field is the determination of the best nature-inspired metaheuristics for this problem. Even though the "no free lunch" theorem has shattered the notion of finding a universal "best method" for all optimization problems [62], this result is not as limiting as it could seem at first sight, because it talks in terms of average. In this sense, it would still be possible to find a method outperforming any other for a specific given problem. But even if this does not hold, it is reasonable (and also desirable) to compare the performance of different metaheuristics for this reconstruction problem. To this purpose, a first valuable step would be the generation of a reliable, standardized benchmark for the field. Its relevance lies in the fact to provide a common ground for comparative analysis among the different metaheuristics, something that the curve/surface reconstruction research community still lacks. At its turn, this comparative task also requires to address other important issues, such as the parameter tuning. It is well-known that the parameter setting is strongly dependent on the problem, meaning that the set of parameters leading to optimal performance for a given problem might be completely inadequate for other problems. Clearly, further research is expected in this area for coming years.

In addition to the theoretical results to be developed, there is also a promising research line regarding the potential applications of the resulting technology in this field. On one hand, we have already identified several possible practical applications in the fields of geometric modeling and processing and CAD/CAM [4, 63–67], with relevance in problems such as the generation of tool-path trajectories for computer numerical controlled machining and milling. On the other hand, the popularization and widespread dissemination of sensor and capturing technologies such as 3D laser scanners, LIDAR models, or photogrammetry, along with the wide availability of affordable devices for 3D printing will bring a growing trend of mass customization of goods and products in next years. The increasing level of sophistication of customized designs will lead to the development of new methods for shape reconstruction. As a result, we anticipate a new golden era of nature-inspired computational intelligence methods for shape reconstruction in reverse engineering.

7 Conclusions

In this chapter, we claim that the curve/surface data fitting problem for reverse engineering can be successfully solved by following a nature-inspired computational intelligence approach. Our starting point is the conversion of this problem from the purely geometric approach based on capturing the underlying shape of free-form objects to the mathematical formulation of this problem as a nonlinear continuous optimization problem. Furthermore, this problem is generally multimodal and of very high dimension, and exhibits challenging features such as noisy data and irregular sampling, making it hard (if not impossible) to be properly addressed by traditional mathematical optimization techniques. Then, we turn our attention to the use of metaheuristics, with emphasis on three recent nature-inspired swarm intelligence approaches: firefly algorithm, cuckoo search algorithm, and bat algorithm. Some examples of application of these techniques reported in the literature for different cases of curves and surfaces are also briefly described. Finally, we envision some future trends and promising lines of research in the field.

Acknowledgments This research has been kindly supported by the Computer Science National Program of the Spanish Ministry of Economy and Competitiveness, Project Ref. #TIN2012-30768, Toho University (Funabashi, Japan), and the University of Cantabria (Santander, Spain). The authors are particularly grateful to the Department of Information Science of Toho University for all the facilities given to carry out this work. A special recognition is also owe to Prof. Xin-She Yang for his kind assistance during the process of dealing with the methods described in this chapter and the very helpful material he provided us during several stages of this work.

References

1. Barnhill, R.E.: Geometric Processing for Design and Manufacturing. SIAM, Philadelphia (1992)
2. Pottmann, H., Leopoldseder, S., Hofer, M., Steiner, T., Wang, W.: Industrial geometry: recent advances and applications in CAD. Comput. Aided Des. **37**, 751–766 (2005)
3. Farin, G.: Curves and surfaces for CAGD, 5th edn. Morgan Kaufmann, San Francisco (2002)

4. Patrikalakis, N.M., Maekawa, T.: Shape Interrogation for Computer Aided Design and Manu-facturing. Springer, Heidelberg (2002)
5. Dierckx, P.: Curve and Surface Fitting with Splines. Oxford University Press, Oxford (1993)
6. Powell, M.J.D.: Curve fitting by splines in one variable. In: Hayes, J.G. (ed.) Numerical Approx-imation to Functions and Data. Athlone Press, London (1970)
7. Park, H.: An error-bounded approximate method for representing planar curves in B-splines. Comput. Aided Geom. Des. **21**, 479–497 (2004)
8. Wang, W.P., Pottmann, H., Liu, Y.: Fitting B-spline curves to point clouds by curvature-based squared distance minimization. ACM Trans. Graphics **25**(2), 214–238 (2006)
9. Park, H., Lee, J.H.: B-spline curve fitting based on adaptive curve refinement using dominant points. Comput. Aided Des. **39**, 439–451 (2007)
10. Barhak, J., Fischer, A.: Parameterization and reconstruction from 3D scattered points based on neural network and PDE techniques. IEEE Trans. Visual. Comput. Graphics **7**(1), 1–16 (2001)
11. Gu, P., Yan, X.: Neural network approach to the reconstruction of free-form surfaces for reverse engineering. Comput. Aided Des. **27**(1), 59–64 (1995)
12. Hoffmann, M.: Numerical control of Kohonen neural network for scattered data approximation. Numer. Algorithms **39**, 175–186 (2005)
13. Kumar, S.G., Kalra, P.K., Dhande, S.G.: Curve and surface reconstruction from points: an approach based on self-organizing maps. Appl. Soft Comput. **5**(5), 55–66 (2004)
14. Yu, Y.: Surface reconstruction from unorganized points using self-organizing neural networks. Proc. IEEE Vis. **99**, 61–64 (1999)
15. Iglesias, A., Gálvez, A.: A new artificial intelligence paradigm for computer aided geometric design. Lect. Notes Artif. Intell. **2001**, 200–213 (1930)
16. Iglesias, A., Gálvez, A.: Hybrid functional-neural approach for surface reconstruction. Math. Prob. Eng. **351648**, 13 (2014)
17. Echevarría, G., Iglesias, A., Gálvez, A.: Extending neural networks for B-spline surface recon-struction. Lect. Notes Comput. Sci. **2330**, 305–314 (2002)
18. Iglesias, A., Echevarría, G., Gálvez, A.: Functional networks for B-spline surface reconstruc-tion. Future Gener. Comput. Syst. **20**(8), 1337–1353 (2004)
19. Iglesias, A., Gálvez, A.: Curve fitting with RBS functional networks. In: Proceedings of Inter-national Conference on Convergence Information Technology-ICCIT'2008, Busan (Korea), pp. 299–306. IEEE Computer Society Press, Los Alamitos, California (2008)
20. Sarfraz, M., Raza, S.A.: Capturing outline of fonts using genetic algorithms and splines. In: Proceedings of Fifth International Conference on Information Visualization IV'2001, pp. 738–743. IEEE Computer Society Press (2001)
21. Gálvez, A., Iglesias, A., Puig-Pey, J.: Iterative two-step genetic-algorithm method for efficient polynomial B-spline surface reconstruction. Inf. Sci. **182**(1), 56–76 (2012)
22. Yoshimoto, F., Harada, T., Yoshimoto, Y.: Data fitting with a spline using a real-coded algorithm. Comput. Aided Des. **35**, 751–760 (2003)
23. Gálvez, A., Cobo, A., Puig-Pey, J., Iglesias, A.: Particle swarm optimization for Bézier surface reconstruction. Lect. Notes Comput. Sci. **5102**, 116–125 (2008)
24. Gálvez, A., Iglesias, A.: Efficient particle swarm optimization approach for data fitting with free knot B-splines. Comput. Aided Des. **43**(12), 1683–1692 (2011)
25. Gálvez, A., Iglesias, A.: Particle swarm optimization for non-uniform rational B-spline surface reconstruction from clouds of 3D data points. Inf. Sci. **192**(1), 174–192 (2012)
26. Loucera, C., Gálvez, A., Iglesias, A.: Simulated annealing algorithm for Bézier curve approx-imation. In: Proceedings of Cyberworlds 2014, pp. 182–189. IEEE Computer Society Press, Los Alamitos, CA (2014)
27. Jing, L., Sun, L.: Fitting B-spline curves by least squares support vector machines. In: Pro-ceedings of the 2nd International Conference on Neural Networks & Brain, pp. 905–909. IEEE Press, Beijing (China) (2005)
28. Gálvez A., Iglesias A., Avila, A.: Immunological-based approach for accurate fitting of 3D noisy data points with Bézier surfaces. In: Proceedings of the Internation Conference on Computer Science-ICCS'2013. Procedia Computer Science, vol. 18, pp. 50–59 (2013)

29. Gálvez, A., Iglesias, A., Avila, A.: Applying clonal selection theory to data fitting with rational Bézier curves. In: Proceedings of the Cyberworlds 2014. IEEE Computer Society Press. Los Alamitos, CA, 221–228 (2014)
30. Gálvez, A., Iglesias, A., Avila, A., Otero, C., Arias, R., Manchado, C.: Elitist clonal selection algorithm for optimal choice of free knots in B-spline data fitting. Appl. Soft Comput. **26**, 90–106 (2015)
31. Zhao, X., Zhang, C., Yang, B., Li, P.: Adaptive knot adjustment using a GMM-based continuous optimization algorithm in B-spline curve approximation. Comput. Aided Des. **43**, 598–604 (2011)
32. Gálvez, A., Iglesias, A., Cabellos, L.: Tabu search-based method for Bézier curve parameterization. Int. J. Softw. Eng. Appl. **7**(5), 283–296 (2013)
33. Keller, R.E., Banshaf, W., Mehnen, J., Weinert, K: CAD surface reconstruction from digitized 3D point data with a genetic programming/evolution strategy hybrid. In: Advances in Genetic Programming 3, pp. 41–65. MIT Press, Cambridge, MA, USA (1999)
34. Gálvez A., Iglesias A., New memetic self-adaptive firefly algorithm for continuous optimization. Int. J. Bio-Inspired Comput. (in press)
35. Gálvez, A., Iglesias, A.: A new iterative mutually-coupled hybrid GA-PSO approach for curve fitting in manufacturing. Appl. Soft Comput. **13**(3), 1491–1504 (2013)
36. Gálvez, A., Iglesias, A., Cobo, A., Puig-Pey, J., Espinola, J.: Bézier curve and surface fitting of 3D point clouds through genetic algorithms, functional networks and least-squares approximation. Lect. Notes Comput. Sci. **4706**, 680–693 (2007)
37. Engelbretch, A.P.: Fundam. Comput. Swarm Intell. Wiley, Chichester (2005)
38. Kennedy, J., Eberhart, R.C., Shi, Y.: Swarm Intelligence. Morgan Kaufmann Publishers, San Francisco (2001)
39. Mitchell, M.: An Introduction to Genetic Algorithms (Complex Adaptive Systems). MIT Press (1998)
40. Goldberg, D.E.: Genetic Algorithms in Search, Optimization, and Machine Learning. Addison-Wesley (1989)
41. Holland, J.H.: Adaptation in Natural and Artificial Systems. The University of Michigan Press, Ann Arbor (1975)
42. Kennedy, J., Eberhart, R.C.: Particle swarm optimization. In: IEEE International Conference on Neural Networks, pp. 1942–1948. Perth, Australia (1995)
43. Yang, X.S.: Nature-Inspired Metaheuristic Algorithms, 2nd edn. Luniver Press, Frome (2010)
44. Yang, X.S.: Engineering Optimization: An Introduction with Metaheuristic Applications. Wiley, New Jersey (2010)
45. Yang, X.S.: Firefly algorithms for multimodal optimization. Lect. Notes Comput. Sci. **5792**, 169–178 (2009)
46. Yang, X.S.: Firefly algorithm, stochastic test functions and design optimisation. Int. J. Bio-Inspired Comput. **2**(2), 78–84 (2010)
47. Yang, X.S., Deb, S.: Cuckoo search via Lévy flights. In: Proceedings fo the World Congress on Nature & Biologically Inspired Computing (NaBIC), pp. 210–214. IEEE (2009)
48. Yang, X.S., Deb, S.: Engineering optimization by cuckoo search. Int. J. Math. Model. Numer. Optim. **1**(4), 330–343 (2010)
49. Yang, X.S.: A new metaheuristic bat-inspired algorithm. In: Gonzalez, J.R. et al. (eds.) Nature Inspired Cooperative Strategies for Optimization (NISCO 2010). Studies in Computational Intelligence, vol. 284, pp. 65–74. Springer, Berlin (2010)
50. Yang, X.S., Gandomi, A.H.: Bat algorithm: a novel approach for global engineering optimization. Eng. Comput. **29**(5), 464–483 (2012)
51. Yang, X.S.: Bat algorithm: literature review and applications. Int. J. Bio-Inspired Comput. **5**(3), 141–149 (2013)
52. Yang, X.S.: Bat algorithm for multiobjective optimization. Int. J. Bio-Inspired Comput. **3**(5), 267–274 (2011)
53. Gálvez A., Iglesias A.: Firefly algorithm for Bézier curve approximation. In: Proceedings of International Conference on Computational Science and Its Applications—ICCSA'2013. IEEE Computer Society Press, pp. 81–88 (2013)

54. Gálvez A., Iglesias A.: Cuckoo search with Lévy flights for weighted Bayesian energy functional optimization in global-support curve data fitting. Sci. World J. **138760**, 11 (2014)
55. Iglesias, A., Gálvez, A., Collantes, M.: Bat algorithm for curve parameterization in data fitting with polynomial Bézier curves. In: Proceedings of Cyberworlds 2015, pp. 107–114. IEEE Computer Society Press, Los Alamitos, CA (2015)
56. Iglesias, A., Gálvez, A.: Memetic firefly algorithm for data fitting with rational curves. In: Proceedings of Congress on Evolutionary Computation-CEC'2015, Sendai (Japan), pp. 507–514. IEEE CS Press, CA (2015)
57. Iglesias, A., Gálvez, A., Collantes, M.: Global-support rational curve method for data approximation with bat algorithm. In: Proceedings of AIAI'2015, Bayonne (France). IFIP AICT, vol. 458, pp. 191–205 (2015)
58. Gálvez A., Iglesias A.: Firefly algorithm for polynomial Bézier surface parameterization. J. Appl. Math. **237984**, 9 (2013)
59. Iglesias, A., Gálvez, A., Collantes, M.: A bat algorithm for polynomial Bezier surface parameterization from clouds of irregularly sampled data points. In: Proceedings of ICNC2015, Zhangjiajie (China) pp. 1034–1039. IEEE Computer Society Press, Los Alamitos CA (2015)
60. Gálvez A., Iglesias A.: From nonlinear optimization to convex optimization through firefly algorithm and indirect approach with applications to CAD/CAM. Sci. World J. **283919**, 10 (2013)
61. Gálvez A., Iglesias A.: Firefly algorithm for explicit B-Spline curve fitting to data points. Math. Probl. Eng. **528215**, 12 (2013)
62. Wolpert, D.H., Macready, W.G.: No free lunch theorems for optimization. IEEE Trans. Evol. Comput. **1**(1), 67–82 (1997)
63. Gálvez, A., Iglesias, A., Puig-Pey, J.: Computing parallel curves on parametric surfaces. Appl. Math. Model. **38**, 2398–2413 (2014)
64. Puig-Pey, J., Gálvez, A., Iglesias, A.: Helical curves on surfaces for computer-aided geometric design and manufacturing. Lect. Notes Comput. Sci. **3044**, 771–778 (2004)
65. Puig-Pey, J., Gálvez, A., Iglesias, A.: Some applications of scalar and vector fields to geometric processing of surfaces. Comput. Graphics **29**(5), 723–729 (2005)
66. Puig-Pey, J., Gálvez, A., Iglesias, A., Corcuera, P., Rodríguez, J.: Some problems in geometric processing of surfaces. In: Advances in Mathematical and Statistical Modeling (Series: Statistics for Industry and Technology, SIT), pp. 293–304. Birkhauser, Boston (2008)
67. Puig-Pey, J., Gálvez, A., Iglesias, A., Rodríguez, J., Corcuera, P., Gutiérrez, F.: Polar isodistance curves on parametric surfaces. Lect. Notes Comput. Sci. **2330**, 161–170 (2002)

A Novel Fast Optimisation Algorithm Using Differential Evolution Algorithm Optimisation and Meta-Modelling Approach

Yang Liu, Alan Kwan, Yacine Rezgui and Haijiang Li

Abstract Genetic algorithms (GAs), Particle Swarm Optimisation (PSO) and Differential Evolution (DE) have proven to be successful in engineering model cali-bration problems. In real-world model calibration problems, each model evaluation usually requires a large amount of computation time. The optimisation process usu-ally needs to run the numerical model and evaluate the objective function thousands of times before converging to global optima. In this study, a computational frame-work, known as DE-*RF*, is presented for solving computationally expensive cali-bration problems. We have proposed a dynamic meta-modelling approach, in which Random Forest *(RF)* regression model was embedded into a differential evolution optimisation framework to replace time consuming functions or models. We describe the performance of DE and DE-*RF* when applied to a hard optimisation function and a rainfall-runoff model calibration problem. The simulation results suggest that the proposed optimisation framework is able to achieve good solutions as well as pro-vide considerable savings of the function calls with a very small number of actual evaluations, compared to these traditional optimisation algorithms.

Keywords Differential evolution optimisation · Meta-modelling · Random forest regression · Automatic model calibration

1 Introduction

Population based optimisation methods such as genetic algorithm, particle swarm optimisation and differential evolution have emerged as powerful paradigm for global optimisation problems (Holland 1975; Storn and Price 1995; Kennedy and Eberhart 1995) [1–3]. Over the last decades, these approaches have gained significant interest in numerical model calibration problems [4–8]. One of the biggest drawbacks of using these optimisation methods for model calibration, however, is that they require

Y. Liu (✉) · A. Kwan · Y. Rezgui · H. Li
School of Engineering, Cardiff University, Cardiff CF24 3AA, UK
e-mail: LiuY92@cardiff.ac.uk

© Springer International Publishing Switzerland 2016
X.-S. Yang (ed.), *Nature-Inspired Computation in Engineering*,
Studies in Computational Intelligence 637, DOI 10.1007/978-3-319-30235-5_9

a large number of function evaluations to achieve an acceptable solution. For example, Yapo et al. [7] reported that they needed 68,890 model evaluations to calibrate the Sacramento Soil Moisture Accounting model of the National Weather Service River Forecasting System with 13 parameters. For many real-world numerical model calibration problems, due to time and resource constraints, the number of calls of running a simulation model is very limited especially when each evaluation of the quality of solution is very time consuming. Often the time consuming aspect of simulation arises from the time needed to run simulation models. For a coarse hydrodynamic model, each simulation may take upto 1 min to run on a high performance computer. As the spatial and temporal resolution increases, the simulation time also rapidly increases. One can expect the simulation time to increase to several minutes or even hours for full hydrodynamic models. One of the biggest drawbacks of using these population-based optimisation methods for numerical model optimisation, is that they require a large number of function evaluations. Reducing the number of actual evaluations necessary to reach an acceptable solution is thus of major importance. In spite of advances in computer capacity and speed, the enormous computational cost of running complex, high fidelity scientific and engineering simulations makes it impractical to rely exclusively on simulation codes for the purpose of design optimisation [9]. The use of parallel computing offers a remedy to these problems in reducing the overall computational time, but an alternative method of using meta-models to replace computationally expensive model evaluations is hereby suggested. Meta-models may be used in conjunction with parallel computing to provide even higher computational efficiency. The concept of meta-modeling relies heavily on replacing time consuming simulation models with appropriate models that are much faster without significant loss of accuracy. Lately there has been a large development of different types of meta-models in evolutionary algorithms [9–15]. Jin [9] presented an extensive survey of the research on the use of fitness approximation in evolutionary computation. Jin [9] has classified meta-models which are used currently in four categories, namely polynomial models, kriging models, neural networks and support vector machine [16, 17]. Main issues like approximation levels, approximate model management schemes and model construction techniques were reviewed. He stated it was difficult to construct a meta-model that was globally correct due to typically high dimensionality, poor distribution and limited number of training samples. It has also been suggested that an appropriate procedure must be set up to combine meta-models with a numerical model, a technique known as evolution control or model management. In the work by Jin et al. [11], the convergence property of an evolution strategy (ES) with multilayer perceptron (MLP) neural network based fitness evaluation was investigated. Jin et al. [11] stated the number of the controlled individuals should be around 50 % of the population size. In this paper, we propose a new algorithm and solve stability problem using approximate optimisation framework. Time-efficient approximate models can be particularly beneficial for evaluation when an optimisation method is applied. The proposed method in this paper investigates the use of *RF* regression in conjunction with DE in order to accelerate the search process. This work presents an integration of *RF* with DE by proposing evolution control strategies to estimate fitness values of the solutions. We have named this new optimisation DE-*RF*. The algorithm reported here can serve as a benchmark algorithm aimed at those types of expensive optimisation problems. This approach can substantially reduce the number of fitness evaluations on computationally expensive problems without compromising the good search capabilities of DE.

The two different search methods in this paper are used to investigate the optimisation of a hard test optimisation function and a rainfall-runoff model optimisation. In Sect. 2, the differential evolution algorithm is briefly described. In Sect. 3, the proposed optimisation algorithm using Random Forest predictor for solving the expensive optimisation problem is presented. In Sect. 4, the test example is presented that illustrates the principles and implications of the proposed optimisation algorithm and proposed optimisation DE-*RF* is applied to a rainfall-runoff model calibration problem. Finally, conclusions are given in Sect. 5.

2 Differential Evolution

Differential Evolution (DE) is a population based algorithm proposed by Storn and Price [2] for optimisation problems over a continuous domain and it has been extended to solve discrete problems. The main operators that control the evolutionary process are the mutation and selection operators. The standard DE works as follows: for each vector $x_{i,G}, i = 1, 2, \ldots, P$, a trial vector v is generated according to

$$v_{i,G+1} = x_{r_1,G} + F \cdot (x_{r_2,G} - x_{r_3,G}), \tag{1}$$

with $r_1, r_2, r_3 \in [1, P]$, integer and mutually different, $F > 0$, and $r_1 \neq r_2 \neq r_3 \neq i$. F is a real and constant factor which controls the amplification of the differential variation $(x_{r_2,G} - x_{r_3,G})$. In order to increase the diversity of the parameter vectors, the following vector is adopted:

$$u_{i,G+1} = (u_{1i,G+1}, u_{2i,G+1}, \ldots, u_{Di,G+1}) \tag{2}$$

with:

$$u_{ji,G+1} = \begin{cases} v_{ji,G+1}, & \textit{if(rand}(0, 1) \le CR) \textit{ or } j = \textit{rnbr}(i) \\ x_{ji,G}, & \textit{if(rand}(0, 1) > CR) \textit{ and } j \neq \textit{rnbr}(i) \end{cases} \quad j = 1, 2, \ldots, D \tag{3}$$

where D is the problem dimension; and CR is a user-defined crossover rate. F and CR are both generally in the range [0.5, 1.0] [2]. *rnbr(i)* is a randomly chosen index from [1, D]. In order to decide whether the new vector u shall become a population member at generation $G + 1$, it is compared to $x_{i,G}$. If vector u yields a smaller objective function value than $x_{i,G}$, $x_{i,G+1}$ is set to u, otherwise the old value $x_{i,G}$ is retained.

3   A Fast Optimisation Algorithm Using Differential Evolution and Random Forest Predictor

3.1   Fitness Prediction Using Meta-Model

Meta-models have been widely used for design evaluation and optimisation in many engineering applications. In this paper, an approach is outlined for integrating meta-models with a differential evolution algorithm optimisation. For simplicity of presentation, consider the optimisation problem with simple bound constraints given by:

$$\text{Minimize } f(x) \text{ Subject to } x_l \leq x \leq x_u \tag{4}$$

where $x \in \mathfrak{R}^n$ is the vector of calibration variables, and x_l and x_u are the lower and upper bounds, respectively. The prediction of a meta-model output is formulated as:

$$f_p = f(x) + \varepsilon(x) \tag{5}$$

where $\varepsilon(x)$ is the error function. We are concerned here with problems where evaluation of function $f(x)$ is computationally expensive. Figure 1 shows a hypothetical example, minimisation of a 1-D function using fitness function and meta-model, respectively. The solid line denotes the true function, dashed line the approximate function, the black dots the available samples for true function, the squares the corresponding predicted values, the stars the equal values between true fitness and prediction values and the vertical dashed line the error between prediction value and true objective function value. The meta-model will take a shape similar to the original function and provide the most equal points for current population if we can

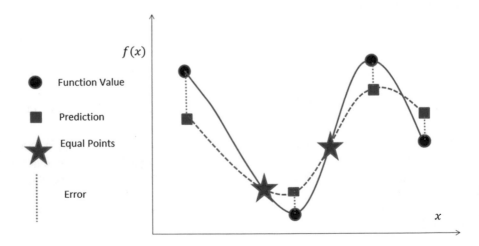

Fig. 1   An example of one-dimensional function for minimization using a meta-model

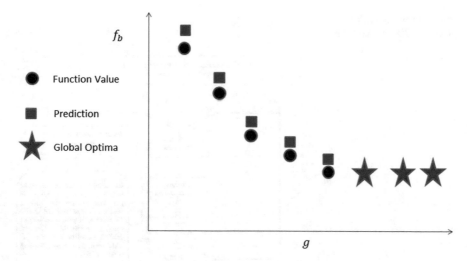

Fig. 2 Iteration process using true objective function and meta-model

obtain perfect global functional approximation of the original fitness function. The corresponding ideal iteration process (the number of iterations g versus best fitness f_b values) using the meta-model is shown in Fig. 2. As the search process continues, the entire population tends to converge to the global optimum (stars) using the meta-model which produces results similar to the true objective function. If the big error exists in the meta-model, it will take a large number of iterations to converge to the global optimum, and it will cause oscillation during optimisation process. In order to solve the unstable problem, dynamically updating training samples to provide accurate prediction during search process and running more iterations than a standard optimisation (e.g. GA, PSO or DE) but with fewer actual evaluations will be proposed. The new control strategy and the dynamic learning approach can guarantee the optimisation using the meta-model to converge towards the global optimum in most cases with fewer actual evaluations (Fig. 3).

3.2 Random Forest

There are many nonlinear classification methods such as artificial neural networks (ANNs), support vector machine (SVM) and RF classifier [16, 17]. An ANN has many neurons and each neuron accepts input and gives output according to its activation function. ANN can learn the mapping relationship between the inputs and the outputs sampled from a training set using a supervised learning algorithm. Then the trained ANN is used to make a prediction for test data. SVM uses a Lagrangian formulation to maximise the margin to separating hyperplane during training process, and it finds a kernel function that maps the inseparable input data to a higher dimensional

Fig. 3 Flow chart of the
proposed DE-*RF* algorithm

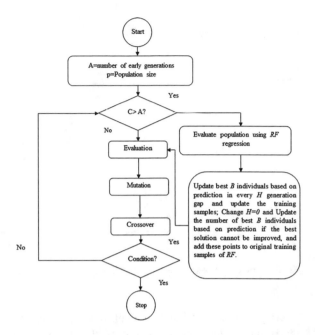

hyperspace. Through this mapping, the training data become more separable, allowing separating hyperplanes to be found to classify the data into similar group. Random
Forest is a regression algorithm that uses an ensemble of unpruned decision trees,
each of which is built on a bootstrap sample of the training data using a randomly
selected subset of variables. RF is an effective tool in prediction because it does not
"overfit" the data. Injecting the right kind of randomness allow them to be accurate
classifiers and regressors [18]. Furthermore, in terms of strength of the individual
predictors and their correlations the framework gives insight into the ability of the
random forest to predict. Breiman [18] found that using out-of-bag estimation makes
concrete the otherwise theoretical values of strength and correlation. In his test for
eight different data sets, RF was always better than bagging and was also competitive
compared to adaptive bagging. We have employed the high-quality implementation
of RF available in the Fortran package "Random Forest". Following the suggestions,
we applied RFs with different parameter configurations for the values of *ntree* =
overall number of trees in the forest, and *mtry Factor* = number of randomly pres-
elected predictor variables for each split. *mtry* is suggested to be square root of the
number of variables [19]. RFs with *ntree* and *mtry Factor* parameters also can be
optimised by cross-validation or optimisation algorithms such as grid search method
and genetic algorithm. Some general remarks can be made about suitable parameter
settings from experience with different linear or non-linear predictors, but it is difficult to provide explicit rules on model selection [4]. Firstly, it is recommended to
implement a simple linear model for a given problem. If a simple model is found to
under-fit the samples, a model with higher complexity should be considered, such

as random forest or support vector machine models. For a more detailed description of random forest and support vector machines the reader is referred to Bishop [16], Schölkopf et al. [17] and Breiman [18].

3.3 Algorithm Design

The goal of our hybrid model is to reduce the number of evaluations of the exact objective functions without losing performance on achieving an acceptable solution. The DE-*RF* algorithm is presented below:

(1) **Initialise**: Select A, P and G, where A = number of early iterations, P = population size, and G = total number of iterations.

(2) **Generate training samples**: Run the DE for A number of generations of population size, P. Compute the initial training samples size S = P × A.

(3) **Construct hybrid-model**: If the current iteration $C > A$, use S as training samples of *RF* predictor; otherwise, run standard DE's process and go to step (8).

(4) **Predict fitness value**: Assign a prediction fitness value to each solution using *RF* model instead of running simulation model after *RF* training process.

(5) **Do DE operations**: Perform the DE operations.

(6) **Update training samples**: Update the number of best B individuals based on prediction in every generation gap H and Use S = S + B points as new training samples of *RF* regression model.

(7) **Change the updated solutions**: If the best solution in training samples cannot be improved on in the continued generations, we switch the generation gap H to 0 or normal DE process and only update the number of best B individuals based on prediction. Here, we weigh up the balance between the prediction accuracy and the number of actual evaluations, so we only update partial population. It has been shown that this strategy to update the best predicted solutions can reduce the actual number further compared to random selection from current predicted population [9]. This is because the best solutions are more important than weak solutions during DE process.

(8) **Check convergence**: Repeat the steps (3)–(8) until the stopping criterion is met.

In order to reduce the number of actual evaluations as much as possible without much loss of accuracy, a new evolution control strategy is hereby proposed: a DE-*RF* that runs more generations than a standard DE but with fewer actual evaluations. The new evolution control strategy should ensure that the DE converges towards to global optimum with a small number of actual evaluations of original function or numerical model. The new evolution control strategy and the dynamic learning approach can guarantee that the DE-*RF* converge towards the global optimum in most cases with fewer actual evaluations. The main advantage using the hybrid optimisation algorithm is thus that DE-*RF* is computational less expensive compared to DE algorithm. It is worth mentioning here that computational time taken to test a *RF* can be ignored compared to the overall computational time taken to run the expensive simulation model. This is because it is assumed in our study that computational time for calling model evaluations is so large, and the size of training samples is so small, that the time taken for *RF* testing will be comparatively small.

4 Application Example

4.1 Bukin Test Function

Bukin function is almost fractal (with fine seesaw edges) in the surroundings of their minimal points (see Fig. 4). Due to this property, it is very difficult to optimize by any method of global (or local) optimisation [20]. In the search domain $x_1 \in [-15, -5]$, $x_2 \in [-3, 3]$, the function is defined as follows.

$$f(x_1, x_2) = 100 \times \sqrt{\left| x_2 - 0.01 \times x_1^2 \right|} + 0.01 \times |x_1 + 10| \tag{6}$$

$$f_{\min}(-10, 1) = 0 \tag{7}$$

4.2 Experimental Setup and Analysis

The relevant experiment parameters using the DE and DE-*RF* for the test function are listed in Tables 1 and 2. For each test, the number of generations equal to 500

Fig. 4 Bukin function

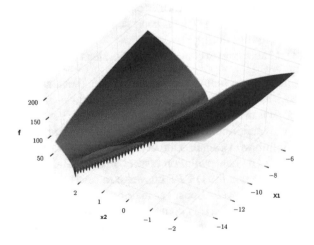

Table 1 Experimental parameters using DE

Parameter	Description	Range
CR	Crossover rate	0.9
F	Control parameter	0.5
G	The total iterations	500
P	Population size	30

Table 2 Experimental parameters using DE-*RF*

Parameter	Description	Recommended value
CR	Crossover rate	0.9
F	Control parameter	0.5
P	Population size	30
H	Generation gap	2
B	The number of controlled individuals	5
A	The number of early generations (Initial training sample set size = P × A)	30
G	The total iterations or generations	600
ntree	The number of trees of Random Forest	200
mtry	The number of randomly preselected predictor variables for each split of Random Forest	1

was employed as a stopping criterion using DE when a population of p = 30 was used. In all the following examples, we report the results obtained from performing 10 random runs of each algorithm compared. G is the total number of generations, H is the generation gap, *B* is the number of controlled individuals when H is changed to 0 during DE optimisation process, and A is the number of early generations.

All the results were averaged over the 10 random runs. The results, including average fitness value and the standard deviations using the two optimisation algorithms for the test objective function, are listed in Tables 3 and 4. It can be seen that the average performance of DE-*RF* is better with respect to the Root Mean Square of the objective function value and has the better stability with a smaller STD value in the 10 random runs compared to traditional DE algorithm. These results show he number of function evaluations using DE-*RF* is around 61 % less than DE. The test results clearly show that the DE-*RF* is capable of converging to acceptable solution with a significantly lower computational budget. When comparing the two different optimisation algorithms, efficiency is a general aspect to consider. In the test case, the DE method used 15,000 model evaluations whereas the DE-*RF* was much more efficient and required fewer function evaluations (about 5702 evaluations). This implies that we get a considerable advantage by using DE-*RF*. The small standard deviations of fitness by the methods DE-*RF* and DE imply that both methods are stable. This outcome may be easily explained. During early iterations

Table 3 Optimisation result using DE

Trial	Optimal function value using DE	Optimal parameter x_1	Optimal parameter x_2
1	0.3410	−14.6725	2.1528
2	0.0208	−12.0617	1.4548
3	0.0316	−11.2557	1.2669
4	1.0694	−8.4232	0.7094
5	0.5361	−11.8694	1.4088
6	0.0709	−6.8343	0.4671
7	0.0160	−10.9027	1.1887
8	0.2507	−14.3691	2.0647
9	0.0332	−8.0460	0.6474
10	0.2362	−13.7181	1.8819
Mean	0.2606		
STD	0.3324		

Table 4 Optimisation result using DE-*RF*

Trial	The number of function evaluations	Optimal function value	Optimal parameter x_1	Optimal parameter x_2
1	5550	0.0045	−10.4447	1.0909
2	5800	0.0041	−9.6002	0.9216
3	5525	0.0041	−10.4091	1.0835
4	6100	0.0144	−11.3977	1.2991
5	5550	0.0172	−8.2836	0.6862
6	5375	0.0158	−8.4152	0.7082
7	5700	0.0124	−11.2102	1.2567
8	6050	0.0027	−9.7369	0.9481
9	5600	0.0027	−10.2590	1.0525
10	5775	0.0067	−9.3402	0.8724
Mean	5702	0.0085		
STD		0.0058		

the two algorithms are identical since they use the same global model for predictions. When training samples of *RF* are dynamically updated, the hybrid model will take a shape similar to the true objective function and provide more prediction accuracy for the current population. This process can guide or correct DE's search in the right direction. Hence, performance studies of DE-*RF* on the objective function reflect the dynamic learning of *RF* predictor's ability to escape from poor local optimal and head towards the global optimum. Overall, the results obtained also imply that the DE-*RF* is capable of identifying the better quality individuals in each DE population. At the same time, *RF*'s dynamic learning process enables *RF* predictor to adapt to local regions (particularly those near the optimal solution). The performance of a hybrid model depends on nonlinear predictor used for fitness estimation, the number of reevaluated solutions, the generation gap to update current solutions and training samples, and the test optimisation function or model. The robust predictor of *RF* for fitness estimation and the dynamic and adaptive learning process are the key reasons for the improvements in search quality at a significantly lower computational cost than the standard DE. The test results show that a well sampled *RF* can achieve satisfactory accuracy. The dynamic learning process allows the hybrid model to adapt to the area in which the DE is searching and provide more accuracy. The DE-*RF* optimisation parameters that must be specified by the user are A, H, *ntree* and *B*: respectively, the number of early generations, the generation gap to update current population and *RF*, the number of trees of Random Forest regression model, and the number of re-evaluated solutions. The parameter A determines the initial training set size because the first fitness estimation is after A generations. The training set size is A × P, where P is the population size. To study the sensitivity of the optimisation results to training sample size A × P, we conducted a series of optimisation runs in which all conditions were identical except for the number of early generations. We tested the DE-*RF* with values of A equal to 10, 20, 30, … 60, so the corresponding training set sizes are: 300, 600, 900, … 1800. Figure 5 shows DE-*RF* performance for different initial training set size A × P. DE-*RF* is not sensitive to the size of the initial training set from the figure because the training samples of *RF* regression model are in any case dynamically updated and increased after the early generations in the hybrid optimisation framework. To reduce the total number of parameter combinations, we adjusted the parameter *H* and fixed the others in each run. We varied the number of updated solutions as *H* = 2, 3, 4, …8, 20, 40 and 60. An inappropriate generation gap *H* would not reduce the number of function evaluations. The final results for the test objective function are shown in Fig. 6 and indicate that:

Fig. 5 Results on different training size (P × A)

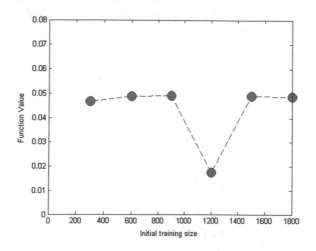

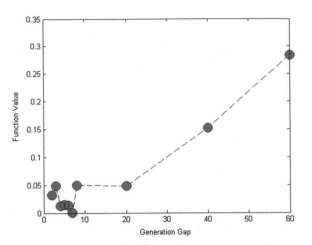

Fig. 6 Results on different generation gap H

(1) the improvement is significant as H decreases from 60 to 20. Hence, DE-*RF* is sensitive to H but the adaptive method to switch two generation gaps (H or 0) can guide or correct DE's search in the right direction with a small value of generation gap H;

(2) the number of actual evaluations will increase as H decreases; and

(3) it is difficult to choose the value of H to balance between performance and the total number of actual evaluations. From these figures, it can be seen that when the parameter H is around 6, the experimental results get closer to global optimum for the test function optimisation with 2 parameters. A large generation gap H to update current predicted population may cause large model error. An adaptive method to switch two generation gaps (H or 0) through monitoring the best solution in the predicated population was suggested in the dynamic learning process. The adaptive method can avoid choosing an improper generation gap for users.

We varied the number of updated solutions as $B = 1, 5, 10, 15$ and 20 when H is switched to 0 value during DE search process. It is shown in Fig. 7 that $B >= 5$ can achieve a good solution for the test objective function. The optimisation performance was not absolutely improved with a large B value because the dynamic and adaptive control of H was applied and the randomness of population was also changed during DE optimization process, and we can see the result from Fig. 7. The predicted accuracy of fitness value using RF mainly depends distribution of two parameters of training data in parameter space. The result from Fig. 7 indicates the dynamic and adaptive control of H can achieve stable result using DE-*RF* without difficulty setting B value. The frequency of H switched to 0 will increase during optimisation process with a small fixed value of B.

To study the sensitivity of the optimisation results to different *ntree* values, we conducted a series of optimisation runs in which all conditions were identical except for *ntree*. We tested the DE-*RF* with values of *ntree* equal to 100, 200, 400, 600,

Fig. 7 Results on different updated solutions B

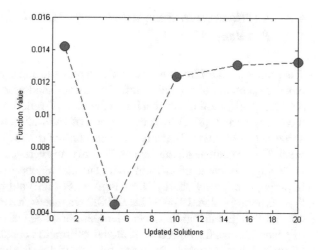

Fig. 8 Results on different *ntree* value of random forest

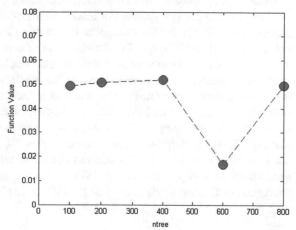

and 800. It is shown in Fig. 8 that *ntree* = 600 is the best value for the test objective function. The *ntree* is not sensitive using DE-*RF* and this can be seen clearly from Fig. 8. A large number of *ntree* will provide a good prediction of fitness value but the prediction accuracy mainly depends distribution of parameters of training data in high dimensional space.

There is only one parameter H to which the performance of DE-*RF* is sensitive and will be decided by the user for other optimisation problems. It is suggested that H takes a relative large value to reduce the number of actual evaluations for other optimisation problems because H is adaptively controlled to provide accurate prediction for current population. The values of the parameter A and P are suggested to be $1 < A \leq 50$ and $20 \leq P \leq 100$, and $H = 3-20$ (generations or iterations), $B = P/2$ (half of population size) and *ntree* = $200-800$ are suggested for other optimisation problems. *mtry* is suggested to be square root of the number of variables [15], and F and CR are both generally in the range [0.5, 1.0] [2].

4.3 Application to Rainfall-Runoff Model Calibration Problem

The proposed algorithm is now tested in a realistic and established engineering optimisation problem of rainfall runoff. The established model used in this study is the NAM rainfall-runoff model that forms the rainfall-runoff module of the MIKE11 river modelling system [8]. A brief description of calibration parameters used is given in Table 5. The MIKE11/ NAM model was applied to the Danish Tryggevaelde catchment. This catchment has an area of $130\,km^2$, an average rainfall of 710 mm/year and an average discharge of 240 mm/year. The catchment is dominated by clayey soils, implying a relatively "flashy" flow regime. For the calibration, historic data from a 5-years period (1 Jan. 1984–31 Dec. 1988) was used where daily data of precipitation, potential evapotranspiration, mean temperature, and catchment runoff are available. In general terms, the objective of model calibration consists of changing the values of model input parameters in an attempt to simulate the hydrological behavior of the catchment as closely as possible. For modeling the rainfall-runoff process at a catchment scale, normally the only available information for evaluating this objective is the catchment runoff. The quality of surface runoff modelling depends on how well it is calibrated. The calibration of the model would be achieved if there is a good match between simulated and observed values of the discharges from the catchment. The root mean squared error (RMSE) between simulated and observed data is used as objective function. For the NAM model calibration using DE, the number of generations equal to 150 is employed as a stopping criterion when a population of $P = 20$ was used. The DE-RF optimisation parameters $P = 20$, $A = 10$, $H = 5$, $ntree = 200$ and $B = 10$ are chosen for calibration of the NAM model. We ran DE-RF for 500 generations as a stopping criterion. With respect to the standard DE and DE-RF, the results in Table 6 clearly show that the DE-RF performed good result as DE at a

Table 5 NAM model parameters

Parameter	Description	Lower limit	Upper limit
Umax (mm)	Maximum water content in the surface storage	5	35
Lmax (mm)	Maximum water content in the lower zone storage	50	350
CQOF	Overland flow runoff coefficient	0	1
CKIF (hour)	Time constant for interflow from the surface storage	500	1000
TIF	Threshold value for interflow	0	0.9
TOF	Threshold value for overland flow	0	0.9
TG	Threshold value for recharge	0	0.9
CK12 (h)	Time constant for overland flow and interflow routing	3	72
CKBF (h)	Baseflow time constant	500	5000

Table 6 Comparison of DE and DE-*RF* for NAM model calibration

Optimisation method	Average number of actual evaluations	Average *F* (RMSE) value	STD	Best	Worst
DE	3000	0.609	0.0035	0.605	0.616
DE-*RF*	1490	0.608	0.003	0.603	0.61

Table 7 Optimal parameter sets (best results) using DE and DE-RF

Parameter	DE	DE-RF
Umax (mm)	9.72	9.85
Lmax (mm)	349.9	349
CQOF	0.743	0.739
CKIF (hour)	501	507
TIF	0.667	0.698
TOF	0.791	0.788
TG	0.726	0.724
CK12 (hour)	34.8	34.5
CKBF (hour)	2225	2307

significantly low computational budget, and both optimisation algorithms are stable with small STD values for 10 random runs. Table 7 shows optimal parameter set by DE and DE-*RF*.

5 Conclusions

In real-world model calibration problems, the model evaluations usually require a large amount of computation time. In this paper, we presented a hybrid optimisation framework (DE-*RF*) that has been developed by combining DE with an approximate *RF* predictor which can together produce fairly accurate global approximations to actual parameters space to provide the function evaluations or model evaluations efficiently. The two optimisation algorithms have been used for optimisation of the test function and the rainfall-runoff model calibration. It is clear that the DE-*RF* approximate framework considered here is capable of searching more efficiently than the standard DE on the objective function, and it does so under a limited computational budget. It is important to choose the proper model values, so that a small number of evaluations will be needed. An adaptive control of the generation gap was suggested and can balance optimisation performance and the small number of actual evaluations. The meta-model we developed can be incorporated with parallel computing technique to improve the performance. Since the hybrid optimisation still uses population concept, the parallel hybrid optimisation performs a distributed computing of allocating a solution that needs to be evaluated using original simulation model or to be estimated using *RF* regression to each processor. This will be investigated in the future.

Acknowledgments The research reported in this paper was conducted as part of the "Developing a Real Time Abstraction & Discharge Permitting Process for Catchment Regulation and Optimised Water Management" project funded by EPSRC (Engineering Physical Sciences Research Council) and TSB (Technology Strategy Board) in the UK as part of the Water Security managed programme (TSB/EPSRC grant reference number: TS/K002805/1). This financial support is gratefully acknowledged.

References

1. Holland, H.J.: Adaptation in Natural and Artificial Systems, An Introductory Analysis with Application to Biology, Control and Artificial Intelligence. The University of Michigan Press, Ann Arbor (1975)
2. Storn, R., Price K.: Differential evolution: a simple and efficient adaptive scheme for global optimisation over continuous spaces. Technical Report TR-95-012, International Computer Science Institute, Berkley (1995)
3. Kennedy, J., Eberhart, R.: Particle Swarm Optimisation. In: Proceedings of the IEEE International Conference on Neural Networks, pp. 1942–1945 (1995)
4. Liu, Y.: Automatic calibration of a Rainfall-Runoff model using a fast and elitist multi-objective particle swarm algorithm. Expert Syst. Appl. **36**(5), 9533–9538 (2009)
5. Liu, Y., Khu, S.T., Savic, D.A.: A fast hybrid optimisation method of multi-objective genetic algorithm and k-nearest neighbour classifier for hydrological model calibration. Lect. Notes Comput. Sci. **3177**, 546–551 (2004)
6. Liu, Y., Pender, G.: Automatic calibration of a rapid flood spreading model using multi-objective optimisations. Soft Comput. **17**, 713–724 (2013)
7. Yapo, P.O., Gupta, H.V., Sorooshian, S.: Multi-objective global optimisation for hydrologic models. J. Hydrol. **204**, 83–97 (1998)
8. Madsen, H.: Automatic calibration of a conceptual rainfall-runoff model using multiple objectives. J. Hydrol. **235**, 276–288 (2000)
9. Jin, Y.: Comprehensive survey of fitness approximation in evolutionary computation. Soft Comput. **9**, 3–12 (2005)
10. Yan, S., Minsker, B.S.: A dynamic meta-model approach to genetic algorithm solution of a risk-based groundwater remediation design model. In: American Society of Civil Engineers (ASCE) Environmental & Water Resources Institute (EWRI) World Water & Environmental Resources Congress 2003 & Related Symposia, Philadelphia, PA (2003)
11. Jin, Y., Olhofer, M., Sendhoff, B.: On evolutionary optimisation with approximate fitness functions. In: Proceedings of the Genetic and Evolutionary Computation Conference (2000)
12. Sun, C.L., Zeng, J.C., Pan, J.Y., Xue, S.D., Jin, Y.C.: A new fitness estimation strategy for particle swarm optimization. Inf. Sci. **221**, 355–370 (2013)
13. Jin, Y.C.: Surrogate-assisted evolutionary computation: recent advances and future challenges. Swarm Evol. Comput. **1**(2), 61–70 (2011)
14. Nguyen, A.T., Reiter, S., Rigo, P.: A review on simulation-based optimization methods applied to building performance analysis. Appl. Energy **113**, 1043–1058 (2014)
15. Forrester, A.I.J., Keane, A.J.: Recent advances in surrogate-based optimization. Prog. Aerosp. Sci. **45**(1–3), 50–79 (2009)
16. Bishop, C.M.: Neural Networks for Pattern Recognition. Oxford University Press (1995)
17. Schölkopf, B., Smola, A., Williamson, R., Bartlett, P.: New support vector algorithms. Neural Comput. **12**, 1207–1245 (2000)

18. Breiman, L.: Random forests. Mach. Learn. **45**(1), 5–32 (2001)
19. Carolin, S., Malley, J., Tutz, G.: Supplement to an introduction to recursive partitioning: rational, application, and characteristics of classification and regression trees, bagging, and random forests (2010). doi:10.1037/a0016973.supp. Accessed 11 Nov 2010
20. Sudhanshu, M.: Some new test functions for global optimization and performance of repulsive particle swarm method. http://www.mpra.ub.uni-muenchen.de/2718/. Accessed 1 Aug 2009

A Hybridisation of Runner-Based and Seed-Based Plant Propagation Algorithms

Muhammad Sulaiman and Abdellah Salhi

Abstract In this chapter we introduce a hybrid plant propagation algorithm which combines the standard PPA which uses runners as a means for search and SbPPA which uses seeds as a means for search. Runners are more suited for exploitation while seeds, when propagated by animals and birds, are more suited for exploration. Combining the two is a natural development to design an effective global optimisation algorithm. PPA and SbPPA will be recalled. The hybrid algorithm is then presented and comparative computational results are reported.

1 Introduction

The initial Plant Propagation Algorithm (PPA), [21], simulates the way strawberry plants propagate. It uses short runners for exploitation and long runners for exploration. Since strawberries propagate using seeds as well as runners, a Seed-based Plant Propagation Algorithm (SbPPA) has also been introduced, [24]. Both algorithms have been shown to be effective on continuous unconstrained and constrained optimisation problems, [24, 25]. However, looking again at the way the strawberry plant propagates, it is easy to see that runners are mainly used to propagate locally (exploitation), while seeds, with the help of birds and animals, are used to propagates further afield (exploration). There is, therefore, sense in considering a hybridisation of PPA and SbPPA, i.e. PPA-SbPPA that fully captures the way the strawberry plant

M. Sulaiman (✉)
Department of Mathematics, Abdul Wali Khan University Mardan,
Mardan, KPK, Pakistan
e-mail: sulaiman513@yahoo.co.uk

A. Salhi
Department of Mathematical Sciences, University of Essex,
Wivenhoe Park, Colchester CO4 3SQ, UK
e-mail: as@essex.ac.uk

© Springer International Publishing Switzerland 2016
X.-S. Yang (ed.), *Nature-Inspired Computation in Engineering*,
Studies in Computational Intelligence 637, DOI 10.1007/978-3-319-30235-5_10

propagates really and to have a truly global search heuristic. This is the subject of this chapter.

The chapter is organised as follows. Section 2 presents PPA. Section 3 presents SbPPA. Section 4 presents PPA-SbPPA. Section 5 reports some comparative experimental results. Section 6 is the conclusion.

2   The Plant Propagation Algorithm or PPA

This is an algorithm that uses short runners to search in the neighbourhood of good local solutions and long runners to search away from poor local solutions, [25]. It starts with an initial population of plants generated using the following equation.

$$x_{i,j} = a_j + (b_j - a_j)\alpha_j, \quad j = 1, \dots, n, \tag{1}$$

where $\alpha_j \in (0, 1)$ is a randomly generated real number for each j. The algorithm proceeds to generate for every member in the population a number n_r of runners; n_r is constant. These runners lead to new solutions, [14], as per the following equation

$$y_{i,j} = x_{i,j} + \beta_j x_{i,j}, \quad j = 1, \dots, n, \tag{2}$$

where $\beta_j \in [-1, 1]$ is a randomly generated number for each j. The term $\beta_j x_{i,j}$ is the length with respect to the jth coordinate of the runner, and $y_{i,j} \in [a_j, b_j]$. If the bounds of the search domain are violated, the point is adjusted to be within the domain.

The generated individual (plant) is evaluated according to the objective function and is stored if it improves the objective function value compared to the mother plant. Otherwise, another plant is created, [14], according to the following equation.

$$y_{i,j} = x_{i,j} + \beta_j b_j, \quad j = 1, \dots, n, \tag{3}$$

where b_j is the jth upper bound and here again $y_{i,j} \in [a_j, b_j]$.

Again, if the newly created plant doesnt improve on the objective value of the mother plant, another one is created, [14], according to equation

$$y_{i,j} = x_{i,j} + \beta_j a_j, \quad j = 1, \dots, n, \tag{4}$$

where a_j is the jth lower bound and $y_{i,j} \in [a_j, b_j]$. This can be considered as a solution at the end of a long runner. Equations 2–4 are repeated in that order if no improvement has been achieved, [14]. A pseudo-code of PPA is as follows.

Algorithm 1 Plant Propagation Algorithm, [25]

1: Create a random population of plants $pop = \{X_i \mid i = 1, 2, \ldots, NP\}$,
   $f \leftarrow$ Objective function , $F \leftarrow$ Temporary population of runners,
2: Evaluate the population,
3: Assume a fixed number of runners as $n_r = 3$,
4: **while** the stopping criteria is not satisfied **do**
5:     Create F;
6:     **for** all plants $i = 1$ to NP **do**
7:         $F_i = X_i$;
8:         **for** $k = 1$ to n_r **do**
9:             Generate a new solution Y according to Equation 2;
10:            Evaluate it and store it in F;
11:            Calculate diff$=\mid f(Y) \mid - \mid f(X_i) \mid$;
12:            **if** diff ≥ 0 **then**
13:                Generate a new solution Y according to Equation 3;
14:                Evaluate it and store it in F;
15:                Compute diff$=\mid f(Y) \mid - \mid f(X_i) \mid$;
16:                **if** diff ≥ 0 **then**
17:                    Generate a new runner using Equation 4;
18:                    Evaluate it and store it in F;
19:                **end if**
20:            **end if**
21:        **end for**
22:    **end for**
23:    Append F to current population pop;
24:    Sort the population in ascending order of the objective values and
       omit the solutions with rank $> NP$;
25:    Update current best;
26: **end while**
27: **return** Best soulution

In the above the following variables/parameters are used. X_i and Y are n-vector solutions; F is a temporary set of created runners which are appended to population pop.

3 The Seed-Based Plant Propagation Algorithm or SbPPA

This algorithm implements a feeding station model to represent the process of animals and birds eating strawberries and then dispersing them, [24]. This in fact is similar to a restaurant opening, waiting for customers which arrive at a ceratin rate, queue and then get served, [7].

We assume that the arrival of birds and animals (dispersing agents) to the plants to feed, is according to the Poisson distribution with mean arrival rate per unit t of time, and population size NP, [18]. They get served (pick up fruit to eat) according to an exponential distribution with mean the average number of agents being served per unit of time t, [3]. We assume that $\lambda < \mu$ and the system in steady state. Basic queuing theory results will give us the expected number of agents and therefore the extent of the dispersal process, [24].

It is important in this algorithm to balance exploration and exploitation, [1, 2]. To this end, we choose a threshold value of the Poisson probability that dictates how much exploration and exploitation is done during the search. The probability $Poiss(\lambda) < 0.05$ means that exploitation is covered. In this case, the equation below is used to help search locally.

$$
x_{i,j}^* = \begin{cases} x_{i,j} + \xi_j(x_{i,j} - x_{l,j}) & \text{if } PR \leq 0.8; \; j = 1, 2, \ldots, n; \\ & i, l = 1, 2, \ldots, NP; \; i \neq l \\ \\ x_{i,j} & Otherwise, \end{cases} \tag{5}
$$

where PR denotes the rate of dispersion of the seeds locally, around the plant; $x_{i,j}^*$ and $x_{i,j} \in [a_j \; b_j]$ are the jth coordinates of the seeds X_i^* and X_i respectively; a_j and b_j are the jth lower and upper bounds defining the search space of the problem and $\xi_j \in [-1 \; 1]$, [20]. The indices l and i are mutually exclusive.

If, on the other hand, $Poiss(\lambda) \geq 0.05$ then the global dispersion of seeds becomes more prominent. This is implemented using the following equation,

$$
x_{i,j}^* = \begin{cases} x_{i,j} + L_i(x_{i,j} - \theta_j) & \text{if } PR \leq 0.8, \; \theta_j \in [a_j \; b_j] \\ & i = 1, 2, \ldots, NP; \; j = 1, 2, \ldots, n \\ \\ x_{i,j} & Otherwise, \end{cases} \tag{6}
$$

where L_i is a step drawn from the Lévy distribution, [26], θ_j is a random coordinate within the search space. Equations 5 and 6 perturb the current solution.

For implementation purposes, we assume that each plant produces one fruit, and each fruit is assumed to have one seed; by a solution X_i we mean the current position of the ith seed to be dispersed. The number of seeds in the population is denoted by NP. Initially we generate a random population of NP seeds using equation

$$
x_{i,j} = a_j + (b_j - a_j)\eta_j, \quad j = 1, \ldots, n, \tag{7}
$$

where $x_{i,j} \in [a_j \; b_j]$ is the jth coordinate of solution X_i, a_j and b_j are the jth coordinates of the bounds describing the search space of the problem and $\eta_j \in (0 \; 1)$. This means that $X_i = [x_{i,j}]$, for $j = 1, \ldots, n$ represents the position of the ith seed in population pop. With the above information, it is possible to describe SbPPA as Algorithm 2.

Algorithm 2 Seed-based Plant Propagation Algorithm (SbPPA), [24]

1: $NP \leftarrow$ Population size, $r \leftarrow$ Counter of trial runs, $MaxExp \leftarrow 30$
2: **for** r=1 : $MaxExp$ **do**
3:     **if** $r \leq NP$ **then**
4:         Create a random population of seeds $pop = \{X_i \mid i = 1, 2, \ldots, NP\}$,
           using Equation 7 and collect the best solutions from each trial run, in pop_{best}.
5:         Evaluate the population pop.
6:     **end if**
7:     **while** $r > NP$ **do**
8:         Use updated population pop_{best}.
9:     **end while**
10:    **while** (the stopping criteria is not satisfied) **do**
11:        **for** $i = 1$ to NP **do**
12:            **if** $Poiss(\lambda)_i \geq 0.05$ **then**,   ▷ (Global or local seed dispersion)
13:                **for** $j = 1$ to n **do**   ▷ (n is number of dimensions)
14:                    **if** rand $\leq PR$ **then**,   ▷ (PR=Perturbation Rate)
15:                        Update the current entry according to Equation 6
16:                    **end if**
17:                **end for**
18:            **else**
19:                **for** $j = 1$ to n **do**
20:                    **if** rand $\leq PR$ **then**,
21:                        Update the current entry according to Equation 5
22:                    **end if**
23:                **end for**
24:            **end if**
25:        **end for**
26:        Update current best
27:    **end while**
28:    ***Return:*** Updated population and global best solution.
29: **end for**

---

4   A Hybridisation of PPA and SbPPA

In PPA, exploitation is more prominent than exploration while in SbPPA it is the other way around. Putting them together, therefore, is quite obvious. All components are already present in the two Algorithms 1 and 2 in terms of the equations that are needed for exploitation and exploration. PPA-SbPPA is a matter of a judicious implementation that calls upon runners for local search, [21], and seeds, [24], for global search.

In each iteration, a plant produces new plants through seeds or runners. The probability $Poiss(\lambda) \geq 0.05$ means that exploration is enhanced, [24]. In this case, Eq. 5 is used, which is helping the algorithm to search. If the quality of new spot X_i^* is better than the quality of its parent location X_i, then the parent location is replaced. Otherwise the new solution is ignored. If $Poiss(\lambda) < 0.05$ then propagation through runners becomes prominent, [24]. The solutions having good quality values will generate short runners and the ones in poor spot will generate a few long runners, [21]. The quality of a spot is calculated according to Eq. 8, [21],

$$f(X) = \frac{f_{max} - f(X)}{f_{max} - f_{min}}, \tag{8}$$

where f_{min} and f_{max} are respectively the minimum and maximum objective function values in the current population. If $f_{max} - f_{min} < \varepsilon$, where ε is a small positive real number, then all positions in the current population are given a quality value of 0.5, [21]. Let $N_i \in (0, 1)$ be the normalized objective function value or fitness value for any X_i, [21], calculated by,

$$N(X) = \frac{1}{2}(\tanh(4f(X) - 2) + 1).$$

Each runner generated has a distance $dx_j^i \in [-1, 1]^n$, calculated by

$$dx_j^i = 2(1 - N_i)(r - 0.5), \quad \text{for } j = 1, \dots, n, \tag{9}$$

where $r \in (0, 1)$ is also randomly generated, [21].

Having calculated the n-dimensional vector dx^i, the new point to explore, $Y_i = [y_{i,j}]$, for $j = 1, \dots, n$, is given by

$$y_{i,j} = x_{i,j} + (b_j - a_j)dx_j^i, \quad \text{for } j = 1, \dots, n, \tag{10}$$

where a_j and b_j are the lower and upper bounds of the search domain respectively. If the bounds of the search domain are violated, the point is adjusted to be within the domain, [14, 21]. With this basic information, the hybrid algorithms PPA-SbPPA can be described as Algorithm 3.

5   Experimental Settings and Discussion

In our experiments we tested PPA-SbPPA against some recently developed algorithms as well as some well established and standard ones. Our set of test problems includes benchmark constrained and unconstrained optimization problems, [8, 19, 22]. The results are compared in terms of statistics (best, worst, mean and standard deviation) for solutions obtained by PPA-SbPPA; SbPPA, [23, 24]; ABC, [15, 16]; PSO, [12]; FF, [10]; HPA, [17] and SSO-C, [8]. The detailed descriptions of these problems are given in the appendices. The significance of the results in Tables 3 and 5 are shown according to the following notations, [13]:

- (+) when PPA-SbPPA is better
- (−) when PPA-SbPPA is worse
- (≈) when the results are approximately the same as those obtained with PPA-SbPPA

Algorithm 3 PPA-SbPPA

1: $NP \leftarrow$ Population size, $r \leftarrow$ Counter of trial runs, $MaxExp \leftarrow 30$
2: **for** $r=1 : MaxExp$ **do**
3:    **if** $r \leq NP$ **then**
4:       Create a random population of plants $pop = \{X_i \mid i = 1, 2, \ldots, NP\}$,
      using Equation 7 and collect the best solutions from each trial run, in pop_{best}.
5:       Evaluate the population pop.
6:    **end if**
7:    **while** $r > NP$ **do**
8:       Use updated population pop_{best}.
9:    **end while**
10:   **while** (the stopping criteria is not satisfied) **do**
11:     **for** $i = 1$ to NP **do**
12:       **if** $Poiss(\lambda)_i \geq 0.05$ **then,**   ▷ (Global or local search)
13:         **for** $j = 1$ to n **do**   ▷ (n is number of dimensions)
14:           Update the current entry according to Equation 5
15:         **end for**
16:       **else**
17:         **for** $j = 1$ to n **do**
18:           Update the current entry according to Equation 10
19:         **end for**
20:       **end if**
21:     **end for**
22:   **end while**
23:   Record the best of this experiment.
24: **end for**

5.1 Parameter Settings

The parameter settings are give in Tables 1 and 2[1]:

6 Conclusion

In this chapter, we have proposed and investigated a hybrid metaheuristic algorithm referred to as PPA-SbPPA. Which captures the overall propagation process of the strawberry plant. We combined the exploitation capability of PPA with the exploration characteristic of SbPPA. Exploitation is performed through sending many short or few long runners in the neighbourhood of the parent, [21]. The distance used by PPA to update the current solution is more prominent in searching the region around the plant, while the search space is covered well through steps generated by SbPPA. The algorithm is compared with SbPPA, [23, 24], and other standard metaheuristics. The recorded evidence shows that this new hybrid produced optimal or near optimal performance in terms of objective value, on all test problems considered (Tables 3, 4 and 5).

---

[1]The symbol "– " denotes an empty cell.

Table 1 Parameters used for each algorithm for solving unconstrained global optimization problems $f_1 - f_{10}$

PSO, [9, 17]	ABC, [15, 17]	HPA, [17]	SbPPA, [23]	PPA-SbPPA
M = 100	SN = 100	Agents = 100	NP = 10	NP = 10
$G_{max} = \frac{(Dimension \times 20,000)}{M}$	$MCN = \frac{(Dimension \times 20,000)}{SN}$	Iteration number $= \frac{(Dimension \times 20,000)}{Agents}$	Iteration number $= \frac{(Dimension \times 20,000)}{NP}$	Iteration number $= \frac{(Dimension \times 20,000)}{NP}$
$c_1 = 2$	MR = 0.8	$c_1 = 2$	PR = 0.8	PR = 0.8
$c_2 = 2$	$limit = \frac{(SN \times dimension)}{2}$	$c_2 = 2$	$Poiss(\lambda) = 0.05$	$Poiss(\lambda) = 0.05$
$W = \frac{(G_{max} - iteration_{index})}{G_{max}}$	–	$limit = \frac{(Agents \times dimension)}{2}$	–	–
–	–	$W = \frac{(Iteration\ number - iteration_{index})}{Iteration\ number}$	–	–

All experiments are repeated 30 times

Table 2 Parameters used for each algorithm for solving constrained optimization problems $f_{11} - f_{18}$

PSO, [12]	ABC, [16]	FF, [10]	SSO-C, [8]	SbPPA, [23]	PPA-SbPPA
M = 250	SN = 40	Fireflies = 25	N = 50	NP = 10	NP = 10
$G_{max} = 300$	MCN = 6000	Iteration number = 2000	Iteration number = 500	Iteration number = 800	Iteration number = 800
$c_1 = 2$	MR = 0.8	q = 1.5	PF = 0.7	PR = 0.8	PR = 0.8
$c_2 = 2$	–	$\alpha = 0.001$	–	$Poiss(\lambda) = 0.05$	$Poiss(\lambda) = 0.05$
Weight factors = 0.9 to 0.4	–	–	–	–	–

All experiments are repeated 30 times

Table 3 Results obtained by PPA-SbPPA, SbPPA, HPA, PSO and ABC

Fun	Dim	Algorithm	Best	Worst	Mean	SD
1	4	ABC	(+) 0.0129	(+) 0.6106	(+) 0.1157	(+) 0.111
		PSO	(+) 6.8991E-08	(+) 0.0045	(+) 0.001	(+) 0.0013
		HPA	(+) 2.0323E-06	(+) 0.0456	(+) 0.009	(+) 0.0122
		SbPPA	(+) 1.08E-07	(+) 7.05E-06	(+) 3.05E-06	(+) 3.14E-06
		PPA-SbPPA	3.6761E-08	1.1146E-07	8.6565E-08	3.8577E-08
2	2	ABC	(+) 1.2452E-08	(+) 8.4415E-06	(+) 1.8978E-06	(+) 1.8537E-06
		PSO	(≈) 0	(≈) 0	(≈) 0	(≈) 0
		HPA	(≈) 0	(≈) 0	(≈) 0	(≈) 0
		SbPPA	(≈) 0	(≈) 0	(≈) 0	(≈) 0
		PPA-SbPPA	0	0	0	0
3	2	ABC	(≈) 0	(+) 4.8555E-06	(+) 4.1307E-07	(+) 1.2260E-06
		PSO	(≈) 0	(+) 3.5733E-07	(+) 1.1911E-08	(+) 6.4142E-08
		HPA	(≈) 0	(≈) 0	(≈) 0	(≈) 0
		SbPPA	(≈) 0	(≈) 0	(≈) 0	(≈) 0
		PPA-SbPPA	0	0	0	0
4	2	ABC	(≈) −1.03163	(≈) −1.03163	(≈) −1.03163	(≈) 0
		PSO	(≈) −1.03163	(≈) −1.03163	(≈) −1.03163	(≈) 0
		HPA	(≈) −1.03163	(≈) −1.03163	(≈) −1.03163	(≈) 0
		SbPPA	(≈) −1.031628	(≈) −1.031628	(≈) −1.031628	(≈) 0
		PPA-SbPPA	−1.031628	−1.031628	−1.031628	0

(continued)

Table 3 (continued)

Fun	Dim	Algorithm	Best	Worst	Mean	SD
5	6	ABC	(≈) −50.0000	(≈) −50.0000	(≈) −50.0000	(−) 0
		PSO	(≈) −50.0000	(≈) −50.0000	(≈) −50.0000	(−) 0
		HPA	(≈) −50.0000	(≈) −50.0000	(≈) −50.0000	(−) 0
		SbPPA	(≈) −50.0000	(≈) −50.0000	(≈) −50.0000	(+) 5.88E-09
		PPA-SbPPA	−50.0000	−50.0000	−50.0000	2.180720E-12
6	10	ABC	(+) −209.9929	(+) −209.8437	(+) −209.9471	(+) 0.044
		PSO	(≈) −210.0000	(−) −210.0000	(−) −210.0000	(−) 0
		HPA	(≈) −210.0000	(−) −210.0000	(−) −210.0000	(+) 1
		SbPPA	(≈) −210.0000	(−) −210.0000	(−) −210.0000	(−) 4.86E-06
		PPA-SbPPA	−210	−209.9999	−209.9999	5.0405E-05
7	30	ABC	(+) 2.6055E-16	(+) 5.5392E-16	(+) 4.7403E-16	(+) 9.2969E-17
		PSO	(≈) 0	(≈) 0	(≈) 0	(≈) 0
		HPA	(≈) 0	(≈) 0	(≈) 0	(≈) 0
		SbPPA	(≈) 0	(≈) 0	(≈) 0	(≈) 0
		PPA-SbPPA	0	0	0	0
8	30	ABC	(+) 2.9407E-16	(+) 5.5463E-16	(+) 4.8909E-16	(+) 9.0442E-17
		PSO	(≈) 0	(≈) 0	(≈) 0	(≈) 0
		HPA	(≈) 0	(≈) 0	(≈) 0	(≈) 0
		SbPPA	(≈) 0	(≈) 0	(≈) 0	(≈) 0
		PPA-SbPPA	0	0	0	0

(continued)

Table 3 (continued)

Fun	Dim	Algorithm	Best	Worst	Mean	SD
9	30	ABC	(≈) 0	(+) 1.1102E-16	(+) 9.2519E-17	(+) 4.1376E-17
		PSO	(≈) 0	(+) 1.1765E-01	(+) 2.0633E-02	(+) 2.3206E-02
		HPA	(≈) 0	(≈) 0	(≈) 0	(≈) 0
		SbPPA	(≈) 0	(≈) 0	(≈) 0	(≈) 0
		PPA-SbPPA	0	0	0	0
10	30	ABC	(+) 2.9310E-14	(+) 3.9968E-14	(+) 3.2744E-14	(+) 2.5094E-15
		PSO	(≈) 7.9936E-15	(+) 1.5099E-14	(≈) 8.5857E-15	(+) 1.8536E-15
		HPA	(≈) 7.9936E-15	(+) 1.5099E-14	(+) 1.1309E-14	(+) 3.54E-15
		SbPPA	(≈) 7.994E-15	(−) 7.99361E-15	(−) 7.994E-15	(+) 7.99361E-15
		PPA-SbPPA	7.9940E-15	1.1546E-14	8.5857E-15	1.4504E-15

All problems considered in this table are unconstrained (see Table 4)

Table 4 Unconstrained global optimization problems used in experiments

Fun	Ftn. Name	D	C	Range	Min	Formulation
f_1	Colville	4	MN	[−10 10]	0	$f(x) = 100(x_1^2 - x_2) + (x_1 - 1)^2 + (x_3 - 1)^2 + 90(x_3^2 - x_4)^2 + 10.1((x_2 - 1)^2 + (x_4 - 1)^2) + 19.8(x_2 - 1)(x_4 - 1)$
f_2	Matyas	2	UN	[−10 10]	0	$f(x) = 0.26(x_1^2 + x_2^2) - 0.48 x_1 x_2$
f_3	Schaffer	2	MN	[−100 100]	0	$f(x) = 0.5 + \dfrac{\sin^2\left(\sqrt{\sum_{i=1}^n x_i^2}\right) - 0.5}{(1+0.001(\sum_{i=1}^n x_i^2))^2}$
f_4	Six Hump Camel Back	2	MN	[−5 5]	−1.03163	$f(x) = 4x_1^2 - 2.1x_1^4 + \frac{1}{3}x_1^6 + x_1 x_2 - 4x_2^2 + 4x_2^4$
f_5	Trid6	6	UN	[−36 36]	−50	$f(x) = \sum_{i=1}^6 (x_i - 1)^2 - \sum_{i=2}^6 x_i x_{i-1}$
f_6	Trid10	10	UN	[−100 100]	−210	$f(x) = \sum_{i=1}^{10} (x_i - 1)^2 - \sum_{i=2}^{10} x_i x_{i-1}$
f_7	Sphere	30	US	[−100 100]	0	$f(x) = \sum_{i=1}^n x_i^2$
f_8	SumSquares	30	US	[−10 10]	0	$f(x) = \sum_{i=1}^n i x_i^2$
f_9	Griewank	30	MN	[−600 600]	0	$f(x) = \frac{1}{4000}\sum_{i=1}^n x_i^2 - \prod_{i=1}^n \cos\left(\frac{x_i}{\sqrt{i}}\right) + 1$
f_{10}	Ackley	30	MN	[−32 32]	0	$f(x) = -20\exp\left(-0.2\sqrt{\frac{1}{n}\sum_{i=1}^n x_i^2}\right) - \exp\left(\frac{1}{n}\sum_{i=1}^n \cos(2\pi x_i)\right) + 20 + e$

Table 5 Results obtained by PPA-SbPPA, SbPPA, PSO, ABC, FF and SSO-C

Fun	Fun name	Optimal	Algorithm	Best	Mean	Worst	SD
11	CP1	−15	PSO	(≈) −15	(≈) −15	(≈) −15	(−) 0
			ABC	(≈) −15	(≈) −15	(≈) −15	(−) 0
			FF	(+) 14.999	(+) 14.988	(+) 14.798	(+) 6.40E-07
			SSO-C	(≈) −15	(≈) −15	(≈) −15	(−) 0
			SbPPA	(≈) −15	(≈) −15	(≈) −15	(+) 1.95E-15
			PPA-SbPPA	−15	−15	−15	1.94E-15
12	CP2	−30665.539	PSO	(≈) −30665.5	(+) −30662.8	(+) −30650.4	(+) 5.20E-02
			ABC	(≈) −30665.5	(+) −30664.9	(+) −30659.1	(+) 8.20E-02
			FF	(≈) −3.07E + 04	(+) −30662	(+) −30649	(+) 5.20E-02
			SSO-C	(≈) −3.07E + 04	(≈) −30665.5	(+) −30665.1	(+) 1.10E-04
			SbPPA	(≈) −30665.5	(≈) −30665.5	(≈) −30665.5	(+) 2.21E-06
			PPA-SbPPA	−3.0665.5	−3.0665.5	−3.0665.5	5.6359E-12
13	CP3	−6961.814	PSO	(+) −6.96E+03	(+) −6958.37	(+) −6942.09	(−) 6.70E-02
			ABC	(−) −6961.81	(+) −6958.02	(+) −6955.34	(−) 2.10E-02
			FF	(+) −6959.99	(+) −6.95E+03	(+) −6947.63	(−) 3.80E-02
			SSO-C	(−) −6961.81	(+) −6961.01	(+) −6960.92	(−) 1.10E-03
			SbPPA	(+) −6961.5	(−) −6961.38	(−) −6961.45	(−) 0.043637
			PPA-SbPPA	−6961.80	−6961.03	−6959.3	0.94896
14	CP4	24.306	PSO	(−) 24.327	(−) 2.45E+01	(+) 24.843	(+) 1.32E-01
			ABC	(−) 24.48	(+) 2.66E+01	(+) 28.4	(+) 1.14
			FF	(+) 23.97	(+) 28.54	(+) 30.14	(+) 2.25
			SSO-C	(−) 24.306	(−) 24.306	(−) 24.306	(−) 4.95E-05
			SbPPA	(−) 24.3444	(−) 24.37536	(−) 24.37021	(−) 0.012632
			PPA-SbPPA	24.529	24.601	24.66	5.3527E-02

(continued)

Table 5 (continued)

Fun	Fun Name	Optimal	Algorithm	Best	Mean	Worst	SD
15	CP5	−0.7499	PSO	(≈) −0.7499	(+) −0.749	(+) −0.7486	(+) 1.20E-03
			ABC	(≈) −0.7499	(+) −0.7495	(+) −0.749	(+) 1.67E-03
			FF	(+) −0.7497	(+) −0.7491	(+) −0.7479	(+) 1.50E-03
			SSO-C	(≈) −0.7499	(≈) −0.7499	(≈) −0.7499	(-) 4.10E-09
			SbPPA	(≈) 0.7499	(≈) 0.749901	(≈) 0.7499	(-) 1.66E-07
			PPA-SbPPA	0.7499	0.7499	0.75	3.25E-06
16	Spring Design Problem	Not Known	PSO	(+) 0.012858	(+) 0.014863	(+) 0.019145	(+) 0.001262
			ABC	(≈) 0.012665	(+) 0.012851	(+) 0.01321	(+) 0.000118
			FF	(≈) 0.012665	(+) 0.012931	(+) 0.01342	(+) 0.001454
			SSO-C	(≈) 0.012665	(+) 0.012765	(+) 0.012868	(+) 9.29E-05
			SbPPA	(≈) 0.012665	(≈) 0.012666	(≈) 0.012666	(-) 3.39E-10
			PPA-SbPPA	0.012665	0.012666	0.012666	4.4532E-08
17	Welded Beam Design Problem	Not Known	PSO	(+) 1.846408	(+) 2.011146	(+) 2.237389	(+) 0.108513
			ABC	(+) 1.798173	(+) 2.167358	(+) 2.887044	(+) 0.254266
			FF	(≈) 1.72485	(+) 2.197401	(+) 2.931001	(+) 0.195264
			SSO-C	(≈) 1.72485	(+) 1.746462	(+) 1.799332	(+) 0.02573
			SbPPA	(≈) 1.72485	(≈) 1.72485	(-) 1.724852	(-) 4.06E-08
			PPA-SbPPA	1.72485	1.72485	1.72487	8.24E-06
18	Speed Reducer Design Optimization	Not Known	PSO	(+) 3044.453	(+) 3079.262	(+) 3177.515	(+) 26.21731
			ABC	(+) 2996.116	(+) 2998.063	(+) 3002.756	(+) 6.354562
			FF	(+) 2996.947	(+) 3000.005	(+) 3005.836	(+) 8.356535
			SSO-C	(≈) 2996.113	(≈) 2996.113	(≈) 2996.113	(+) 1.34E-12
			SbPPA	(≈) 2996.114	(≈) 2996.114	(≈) 2996.114	(≈) 0
			PPA-SbPPA	2996.114	2996.114	2996.114	0

All problems considered in this table are standard constrained optimization problems (see Appendix)

Appendix

Constrained Global Optimization Problems

CP1

$$\text{Min} \quad f(x) = 5\sum_{d=1}^{4} x_d - 5\sum_{d=1}^{4} x_d^2 - \sum_{d=5}^{13} x_d,$$
$$\text{subject to } g_1(x) = 2x_1 + 2x_2 + x_{10} + x_{11} - 10 \le 0,$$
$$g_2(x) = 2x_1 + 2x_3 + x_{10} + x_{12} - 10 \le 0,$$
$$g_3(x) = 2x_2 + 2x_3 + x_{11} + x_{12} - 10 \le 0,$$
$$g_4(x) = -8x_1 + x_{10} \le 0,$$
$$g_5(x) = -8x_2 + x_{11} \le 0,$$
$$g_6(x) = -8x_3 + x_{12} \le 0,$$
$$g_7(x) = -2x_4 - x_5 + x_{10} \le 0,$$
$$g_8(x) = -2x_6 - x_7 + x_{11} \le 0,$$
$$g_9(x) = -2x_8 - x_9 + x_{12} \le 0,$$

where bounds are $0 \le x_i \le 1$ $(i = 1, \ldots, 9, 13)$, $0 \le x_i \le 100$ $(i = 10, 11, 12)$. The global optimum is at $x^* = (1, 1, 1, 1, 1, 1, 1, 1, 1, , 3, 3, 3, 1)$, $f(x^*) = -15$.

CP2

$$\text{Min} \quad f(x) = 5.3578547x_2 + 0.8356891x_1x_5 + 37.293239x_1 - 40792.141,$$
$$\text{subject to } g_1(x) = 85.334407 + 0.0056858x_2x_5 + 0.0006262x_1x_4 - 0.0022053x_3x_5 - 92 \le 0,$$
$$g_2(x) = -85.334407 - 0.0056858x_2x_5 - 0.0006262x_1x_4 + 0.0022053x_3x_5 \le 0,$$
$$g_3(x) = 80.51249 + 0.0071317x2x5 + 0.0029955x1x2 - 0.0021813x_2 - 110 \le 0,$$
$$g_4(x) = -80.51249 - 0.0071317x_2x_5 + 0.0029955x_1x_2 - 0.0021813x_2 + 90 \le 0,$$
$$g_5(x) = 9.300961 - 0.0047026x_3x_5 - 0.0012547x_1x_3 - 0.0019085x_3x_4 - 25 \le 0,$$
$$g_6(x) = -9.300961 - 0.0047026x_3x_5 - 0.0012547x_1x_3 - 0.0019085x_3x_4 + 20 \le 0,$$

where $78 \le x_1 \le 102$, $33 \le x_2 \le 45$, $27 \le x_i \le 45$ $(i = 3, 4, 5)$. The optimum solution is $x^* = (78, 33, 29.995256025682, 45, 36.775812905788)$, where $f(x^*) = -30665.539$. Constraints g_1 and g_6 are active.

CP3

$$\text{Min} \quad f(x) = (x_1 - 10)^3 + (x_2 - 20)^3,$$
$$\text{subject to} \quad g_1(x) = -(x_1 - 5)^2 - (x_2 - 5)^2 + 100 \leq 0,$$
$$g_2(x) = (x_1 - 6)^2 + (x^2 - 5)^2 - 82.81 \leq 0,$$

where $13 \leq x_1 \leq 100$ and $0 \leq x_2 \leq 100$. The optimum solution is $x^* = (14.095, 0.84296)$ where $f(x^*) = -6961.81388$. Both constraints are active.

CP4

$$\text{Min} \quad f(x) = x_1^2 + x_2^2 + x_1 x_2 - 14x_1 - 16x_2 + (x_3 - 10)^2 + 4(x_4 - 5)^2 + (x_5 - 3)^2$$
$$+ 2(x_6 - 1)^2 + 5x_7^2 + 7(x_8 - 11)^2 + 2(x_9 - 10)^2 + (x_{10} - 7)^2 + 45,$$
$$\text{subject to} \quad g_1(x) = -105 + 4x_1 + 5x_2 - 3x_7 + 9x_8 \leq 0,$$
$$g_2(x) = 10x_1 - 8x_2 - 17x_7 + 2x_8 \leq 0,$$
$$g_3(x) = -8x_1 + 2x_2 + 5x_9 - 2x_{10} - 12 \leq 0,$$
$$g_4(x) = 3(x_1 - 2)^2 + 4(x_2 - 3)^2 + 2x_3^2 - 7x_4 - 120 \leq 0,$$
$$g_5(x) = 5x_1^2 + 8x_2 + (x_3 - 6)^2 - 2x_4 - 40 \leq 0,$$
$$g_6(x) = x_1^2 + 2(x_2 - 2)^2 - 2x_1 x_2 + 14x_5 - 6x_6 \leq 0,$$
$$g_7(x) = 0.5(x_1 - 8)^2 + 2(x_2 - 4)^2 + 3x_5^2 - x_6 - 30 \leq 0,$$
$$g_8(x) = -3x_1 + 6x_2 + 12(x_9 - 8)^2 - 7x_{10} \leq 0,$$

where $-10 \leq x_i \leq 10$ $(i = 1, \ldots, 10)$. The global optimum is $x^* = (2.171996, 2.363683, 8.773926, 5.095984, 0.9906548, 1.430574, 1.321644, 9.828726, 8.280092, 8.375927)$, where $f(x^*) = 24.3062091$. Constraints g_1, g_2, g_3, g_4, g_5 and g_6 are active.

CP5

$$\text{Min} \quad f(x) = x_1^2 + (x_2 - 1)^2$$
$$\text{subject to} \quad g_1(x) = x_2 - x_1^2 = 0,$$

where $1 \leq x_1 \leq 1, 1 \leq x_2 \leq 1$. The optimum solution is $x^* = (\pm 1/\sqrt{(2)}, 1/2)$, where $f(x^*) = 0.7499$.

Welded Beam Design Optimisation

The welded beam design is a standard test problem for constrained design optimisation, [6, 27]. There are four design variables: the width w and length L of the welded area, the depth d and thickness h of the main beam. The objective is to minimise the overall fabrication cost, under the appropriate constraints of shear stress τ, bending stress σ, buckling load P and maximum end deflection δ. The optimization model is summarized as follows, where $x^T = (w, L, d, h)$.

$$Minimise \quad f(x) = 1.10471w^2 L + 0.04811dh(14.0 + L), \qquad (11)$$

subject to

$$
\begin{aligned}
g_1(x) &= w - h \leq 0, \\
g_2(x) &= \delta(x) - 0.25 \leq 0, \\
g_3(x) &= \tau(x) - 13,600 \leq 0, \\
g_4(x) &= \sigma(x) - 30,000 \leq 0, \\
g_5(x) &= 1.10471w^2 + 0.04811dh(14.0 + L) - 5.0 \leq 0, \\
g_6(x) &= 0.125 - w \leq 0, \\
g_7(x) &= 6000 - P(x) \leq 0,
\end{aligned}
\qquad (12)
$$

where

$$\sigma(x) = \frac{504,000}{hd^2},$$

$$D = \frac{1}{2}\sqrt{L^2 + (w + d)^2},$$

$$\delta = \frac{65,856}{30,000hd^3},$$

$$\alpha = \frac{6000}{\sqrt{2}wL},$$

$$P = 0.61423 \times 10^6 \frac{dh^3}{6}\left(1 - \frac{\sqrt[d]{\frac{30}{48}}}{28}\right).$$

$$Q = 6000\left(14 + \frac{L}{2}\right),$$

$$J = \sqrt{2}wL\left(\frac{L^2}{6} + \frac{(w + d)^2}{2}\right),$$

$$\beta = \frac{QD}{J},$$

$$\tau(x) = \sqrt{\alpha^2 + \frac{\alpha\beta L}{D} + \beta^2},$$

$$(13)$$

The simple limit or bounds are $0.1 \leq L, d \leq 10$ and $0.1 \leq w, h \leq 2.0$.

Spring Design Optimisation

The main objective of this problem, [4, 5] is to minimize the weight of a tension/compression string, subject to constraints of minimum deflection, shear stress, surge frequency, and limits on outside diameter and on design variables. There are three design variables: the wire diameter x_1, the mean coil diameter x_2, and the number of active coils x_3, [6]. The mathematical formulation of this problem, where $x^T = (x_1, x_2, x_3)$, is as follows.

$$Minimise \quad f(x) = (x_3 + 2)x_2 x_1^2, \tag{14}$$

subject to

$$g_1(x) = 1 - \frac{x_2^3 x_3}{7,178 x_1^4} \le 0,$$

$$g_2(x) = \frac{4x_2^2 - x_1 x_2}{12,566(x_2 x_1^3) - x_1^4} + \frac{1}{5,108 x_1^2} - 1 \le 0,$$

$$g_3(x) = 1 - \frac{140.45 x_1}{x_2^2 x_3} \le 0, \tag{15}$$

$$g_4(x) = \frac{x_2 + x_1}{1.5} - 1 \le 0,$$

The simple limits on the design variables are $0.05 \le x_1 \le 2.0$, $0.25 \le x_2 \le 1.3$ and $2.0 \le x_3 \le 15.0$.

Speed Reducer Design Optimization

The problem of designing speed reducer, [11], is a standard test problem, it consists of the design variables as: face width x_1, module of teeth x_2, number of teeth on pinion x_3, length of the first shaft between bearings x_4, length of the second shaft between bearings x_5, diameter of the first shaft x_6, and diameter of the first shaft x_7 (all variables continuous except x_3 that is integer). The weight of the speed reducer is to be minimized subject to constraints on bending stress of the gear teeth, surface stress, transverse deflections of the shafts and stresses in the shaft, [6]. The mathematical formulation of the problem, where $x^T = (x_1, x_2, x_3, x_4, x_5, x_6, x_7)$, is as follows.

$$Minimise \quad f(x) = 0.7854 x_1 x_2^2 (3.3333 x_3^2 + 14.9334 x_3 43.0934)$$
$$- 1.508 x_1 (x_6^2 + x_7^2) + 7.4777(x_6^3 + x_7^3) + 0.7854(x_4 x_6^2 + x_5 x_7^2), \tag{16}$$

subject to

$$g_1(x) = \frac{27}{x_1 x_2^2 x_3} - 1 \le 0,$$

$$g_2(x) = \frac{397.5}{x_1 x_2^2 x_3^2} - 1 \le 0,$$

$$g_3(x) = \frac{1.93 x_4^3}{x_2 x_3 x_6^4} - 1 \le 0,$$

$$g_4(x) = \frac{1.93 x_5^3}{x_2 x_3 x_7^4} - 1 \le 0,$$

$$g_5(x) = \frac{1.0}{110 x_6^3} \sqrt{\left(\frac{745.0 x_4}{x_2 x_3}\right)^2 + 16.9 \times 10^6} - 1 \le 0,$$

$$g_6(x) = \frac{1.0}{85 x_7^3} \sqrt{\left(\frac{745.0 x_5}{x_2 x_3}\right)^2 + 157.5 \times 10^6} - 1 \le 0,$$ \hfill (17)

$$g_7(x) = \frac{x_2 x_3}{40} - 1 \le 0,$$

$$g_8(x) = \frac{5 x_2}{x_1} - 1 \le 0,$$

$$g_9(x) = \frac{x_1}{12 x_2} - 1 \le 0,$$

$$g_{10}(x) = \frac{1.5 x_6 + 1.9}{x_4} - 1 \le 0,$$

$$g_{11}(x) = \frac{1.1 x_7 + 1.9}{x_5} - 1 \le 0,$$

The simple limits on the design variables are $2.6 \le x_1 \le 3.6$, $0.7 \le x_2 \le 0.8$, $17 \le x_3 \le 28$, $7.3 \le x_4 \le 8.3$, $7.8 \le x_5 \le 8.3$, $2.9 \le x_6 \le 3.9$ and $5.0 \le x_7 \le 5.5$.

References

1. Akay, B., Karaboga, D.: A modified artificial bee colony algorithm for real-parameter optimization. Information Sciences (2010)
2. Akay, B., Karaboga, D.: Artificial bee colony algorithm for large-scale problems and engineering design optimization. J. Intell. Manuf. 23(4), 1001–1014 (2012)
3. Ang, A.H., Tang, W.H.: Probability concepts in engineering. Planning 1(4), 1–3 (2004)
4. Arora, J.: Introduction to optimum design. Academic Press (2004)
5. Belegundu, A.D., Arora, J.: A study of mathematical programming methods for structural optimization. Part I: Theory. Int. J. Numer. Methods Eng. 21(9), 1583–1599 (1985)
6. Cagnina, L.C., Esquivel, S.C., Coello, C.A.C.: Solving engineering optimization problems with the simple constrained particle swarm optimizer. Informatica (Slovenia) 32(3), 319–326 (2008)
7. Cooper, R.B.: Introduction to queueing theory (1972)
8. Cuevas, E., Cienfuegos, M.: A new algorithm inspired in the behavior of the social-spider for constrained optimization. Expert Syst. Appl. Int. J. 41(2), 412–425 (2014)

9. Eberhart, R., Kennedy, J.: A new optimizer using particle swarm theory. In: Micro machine and human science, 1995. MHS'95., Proceedings of the Sixth International Symposium on micro machine and human science, pp. 39–43. IEEE (1995)
10. Gandomi, A.H., Yang, X.S., Alavi, A.H.: Mixed variable structural optimization using firefly algorithm. Comput. Struct. **89**(23), 2325–2336 (2011)
11. Golinski, J.: An adaptive optimization system applied to machine synthesis. Mech. Mach. Theory **8**(4), 419–436 (1974)
12. He, Q., Wang, L.: A hybrid particle swarm optimization with a feasibility-based rule for constrained optimization. Appl. Math. Comput. **186**(2), 1407–1422 (2007)
13. Hong-Yuan, W., Xiu-Jie, D., Qi-Cai, C., Fu-Hua, C.: An improved isomap for visualization and classification of multiple manifolds. In: Lee, M., Hirose, A., Hou, Z.G., Kil, R. (eds.) Neural Information Processing, Lecture Notes in Computer Science, vol. 8227, pp. 1–12. Springer, Berlin Heidelberg (2013). doi:10.1007/978-3-642-42042-9_1. http://dx.doi.org/10.1007/978-3-642-42042-9_1
14. Hooke, R., Jeeves, F.: Direct search solution of numerical and statistical problems. J. Assoc. Comput. Mach. **8** (1961)
15. Karaboga, D.: An idea based on honey bee swarm for numerical optimization. Techn. Rep. TR06, Erciyes Univ. Press, Erciyes (2005)
16. Karaboga, D., Akay, B.: A modified artificial bee colony (abc) algorithm for constrained optimization problems. Appl. Soft Comput. **11**(3), 3021–3031 (2011)
17. Kıran, M.S., Gündüz, M.: A recombination-based hybridization of particle swarm optimization and artificial bee colony algorithm for continuous optimization problems. Appl. Soft Comput. **13**(4), 2188–2203 (2013)
18. Lawrence, J.A., Pasternack, B.A.: Applied management science. Wiley New York (2002)
19. Liang, J., Runarsson, T.P., Mezura-Montes, E., Clerc, M., Suganthan, P., Coello, C.C., Deb, K.: Problem definitions and evaluation criteria for the cec 2006 special session on constrained real-parameter optimization. J. Appl. Mech. **41** (2006)
20. Rahnamayan, S., Tizhoosh, H., Salama, M.: Opposition-based differential evolution. IEEE Trans. Evol. Comput. **12**(1), 64–79 (2008)
21. Salhi, A., Fraga, E.: Nature-inspired optimisation approaches and the new plant propagation algorithm. In: Proceedings of the The International Conference on Numerical Analysis and Optimization (ICeMATH '11), Yogyakarta, Indonesia pp. K2-1–K2-8 (2011)
22. Suganthan, P., Hansen, N., Liang, J., Deb, K., Chen, Y., Auger, A., Tiwari, S.: Problem definitions and evaluation criteria for the cec 2005 special session on real-parameter optimization. Nanyang Technological University, Singapore, Tech. Rep 2005005 (2005)
23. Sulaiman, M., Salhi, A.: The 5th international conference on metaheuristics and nature inspired computing, Morocco. http://meta2014.sciencesconf.org/40158 (2014)
24. Sulaiman, M., Salhi, A.: A seed-based plant propagation algorithm: the feeding station model. Sci. World J. (2015)
25. Sulaiman, M., Salhi, A., Selamoglu, B.I., Kirikchi, O.B.: A plant propagation algorithm for constrained engineering optimisation problems. Math. Probl. Eng. 627416, 10 pp. (2014) doi:10.1155/2014/627416
26. Yang, X.S.: Nature-inspired Metaheuristic Algorithms. Luniver Press (2011)
27. Yang, X.S., Deb, S.: Engineering optimisation by cuckoo search. Int. J. Math. Model. Numer. Optim. **1**(4), 330–343 (2010)

Economic Load Dispatch Using Hybrid MpBBO-SQP Algorithm

Ali R. Al-Roomi and Mohamed E. El-Hawary

Abstract Solving economic load dispatch (ELD) problems of electric power systems means finding the optimal output of the generating units, so that the load demand can be satisfied with the lowest possible operating cost, and without violating any design constraint. Recently, many nature-inspired algorithms have been successfully used to solve these highly constrained non-linear and non-convex ELD problems without facing much efforts as regularly happens with the conventional optimization algorithms. Biogeography-based optimization (BBO) algorithm is a new population-based evolutionary algorithm (EA). As per the conducted studies in the literature, BBO has good exploitation, but it lacks exploration. In this study, the poor exploration level of BBO is enhanced by hybridizing it with the Metropolis criterion of the simulated annealing (SA) algorithm in order to have more control on the migrated individuals; and hence the first phase of this proposed algorithm is called Metropolis BBO (in short MpBBO). The second hybridization phase is done by combining the strength of the Sequential Quadratic Programming (SQP) algorithm with MpBBO to have a new superior algorithm called MpBBO-SQP, where the best solutions per each generation of MpBBO phase is fine-tuned by SQP phase. The performance of MpBBO-SQP is evaluated using three test cases with five different cooling strategies of SA. The results obtained show that MpBBO-SQP outperforms different BBO models as well as many other competitive algorithms presented in the literature.

1 Introduction

Optimal Economic operation is considered as one of the most important problems in any power system that has to be solved in order to satisfy the required load for the end-users by the lowest production cost, so that the net profit can be maximized.

A.R. Al-Roomi (✉) · M.E. El-Hawary
Electrical and Computer Engineering, Dalhousie University,
1459 Oxford Street, Halifax, NS B3H 4R2, Canada
e-mail: alroomi@al-roomi.org

M.E. El-Hawary
e-mail: ElHawary@dal.ca

© Springer International Publishing Switzerland 2016
X.-S. Yang (ed.), *Nature-Inspired Computation in Engineering*,
Studies in Computational Intelligence 637, DOI 10.1007/978-3-319-30235-5_11

There are two strategies that can be used to operate any electric power system in an economical way: the first one is based on the power generation to meet the power demand with the minimum production cost and it is called economic load dispatch (ELD), and the other is based on the control of the power flow to provide the minimum loss in the delivered power to the end-users [1].

To minimize the objective function of the ELD problem, many traditional and modern optimization algorithms have been used [1–14]. Also, this objective involves many constraints (equality, and inequality *discussed in the next section*) that need first to be satisfied in order to get a feasible solution.

Because traditional optimization techniques are single-point searchers, they are very fast as compared with the modern population-based nature-inspired algorithms, if both are initialized with the same number of iterations. However, there are many limitations to these conventional techniques (*lambda-iteration method, the base point and participation factors method, the gradient method, etc.*), like [9, 12]: converging to non-global solutions, derivative-based approaches, the generating units incremental cost curves are linearized. Practically, this engineering problem considers many constraints (*like: generators ramp rate limits, prohibited operating zones and multiple fuel options*), which make conventional methods insufficient for this type of applications. Although, dynamic programming (DP) is successfully used to solve this problem [2, 4], as it consumes a large amount of CPU time and it may end up trapped into local optimum solutions.

Based on that, more recent optimization techniques[1] (*such as: genetic algorithm (GA), evolutionary programming (EP), differential evolution (DF), simulated annealing (SA), particle swarm optimization (PSO), ant-colony optimization (ACO), artificial bee colony (ABC), bacteria foraging optimization (BFO)* [4, 8, 9, 11–14, 16, 17]) are used to overcome inherent complexities that come with the conventional techniques. Thus there is no need to simplify the design function, determine derivative or even select a good initial/starting point for ensuring searching within the global area of the search space. If penalty functions are selected to deal with the constrained problems, then the optimization algorithm can work independently, where the design function and its objective (*whether minimization or maximization*) can be inserted as a plug-in function within the numerical program.

Biogeography-based optimization (BBO) algorithm was presented by Dan. Simon in 2008 [18], and it is considered as a one of meta-heuristic population-based algorithms that can converge to the location of the global optima. The first appearance of this new evolutionary algorithm (EA) was with the simplified linear migration model, where its performance was evaluated based on 14 benchmark functions, and then tested to solve a real sensor selection problem for aircraft engine health estimation. The BBO algorithm proved itself as a very competitive technique as compared with other EAs [18]. Although it has good exploitation for global optimization, it lacks exploration [19–22]. Some of the inherent weaknesses have been addressed

---

[1]The mechanisms of most algorithms are inspired by nature; or, in a more specific, from biology, physics and chemistry sciences [15].

and solved in [23]. However, the performance of the migration stage is still poor and hence the door is open for conducting more studies.

In this study, a hybrid version between BBO, SA and SQP is proposed to solve ELD problem of 3, 13 and 40 units test systems. Although the classical BBO algorithm outperforms many nature-inspired algorithms given in the literature, it has been found that the proposed MpBBO-SQP algorithm can win in this competition.

This chapter is arranged as follows: Sect. 2 shows the general formulation of ELD problems. The detailed description of the proposed hybrid optimization algorithm is covered in Sect. 3. The obtained numerical results of the three test cases are given in Sect. 4; while Sect. 5 is reserved for some further discussion. Finally, conclusions and suggestions are presented in Sect. 6.

2 General Formulation of ELD Problems

The objective function of any ELD problem can be generally expressed as:

$$OBJ = \sum_{i=1}^{n} C_i\left(P_i\right) \tag{1}$$

where C_i is the fuel-cost which is a function of the active power output variable (P_i) of the ith generating unit, and n is the problem dimension or the total number of units.

The variation of fuel-cost on each generating unit (C_i) can be represented by many ways based on the type of each unit [3]. If the ith unit is a conventional type, then the following quadratic function is commonly used:

$$C_i\left(P_i\right) = \alpha_i + \beta_i P_i + \gamma_i P_i^2 \tag{2}$$

where α_i, β_i, and γ_i are the constants of the ith generating machine, and they are defined as follows [10]:

- α_i: cost of crew's salary, interest and depreciation ($)
- β_i: fuel-cost coefficient ($/MWh)
- γ_i: losses measurements in the system ($/MWh^2)

Equation (2) should be extended if the term of the valve-point loading effects is included [13] as follows:

$$C_i\left(P_i\right) = \alpha_i + \beta_i P_i + \gamma_i P_i^2 + \left|e_i \times \sin\left[f_i \times \left(P_i^{min} - P_i\right)\right]\right| \tag{3}$$

where e_i and f_i are fuel-cost coefficients of the ith generating unit with the valve-point loading effects [16].

Moreover, the cost function should also be modified if some additional specifications (*like emission rates and different fuels*) are considered [3, 11]. Furthermore, for non-conventional generators, the quadratic C_i is replaced with other suitable expressions. For example, a linear C_i could be considered for the wind-generated power [24].

Once the objective function is constructed, then some equality, inequality and side constraints should be satisfied in order to have a feasible solution [1, 6]. Some of these constraints are:

2.1 Generator Active Power Capacity Constraint

Each generating unit has its own lower and upper active power limits. This side constraint can be expressed as:

$$P_i^{min} \leqslant P_i \leqslant P_i^{max} \tag{4}$$

where P_i^{min} and P_i^{max} are respectively the minimum and maximum active power that can be supplied by the ith unit.

2.2 Active Power Balance Constrain

In order to meet the consumers' power requirement, the total generated power must satisfy the demand power as well as the losses in the electrical network. This equality constraint can be expressed as:

$$P_T - P_D - P_L = 0 \tag{5}$$

where P_T is the total generated power that can be calculated by Eq. (6). P_D is the total active load demand, and P_L is the transmission losses that can be calculated using Kron's loss formula shown in Eq. (7).

$$P_T = \sum_{i=1}^{n} P_i \tag{6}$$

$$P_L = \sum_{i=1}^{n} \sum_{j=1}^{n} P_i B_{ij} P_j + \sum_{i=1}^{n} B_{0i} P_i + B_{00} \tag{7}$$

where B_{ij}, B_{0i}, and B_{00} are called loss coefficients or just B-coefficients [7].

2.2.1 Equality to Inequality Constraint Conversion

Satisfying equality constraints through a non-gradient iterative process is very hard. Instead, an acceptable amount of tolerance "$\pm\varepsilon$" can be used. Thus, Eq. (5) is satisfied if:

$$- \varepsilon \leqslant h(P) \leqslant \varepsilon, \quad \text{where } h(P) = P_T - P_D - P_L \tag{8}$$

This equation can be split into two parts of inequality constraints as follows:

$$h(P) - \varepsilon \leqslant 0 \tag{9}$$
$$-h(P) - \varepsilon \leqslant 0 \tag{10}$$

2.2.2 External Gradient-Based Iterative Program

Another approach employs a special sub-algorithm to solve Eq. (5) iteratively. For example, Newton-Raphson technique can be employed just to satisfy this constraint.

2.2.3 Slack Generator

The slack generator technique can also be used to directly satisfy this type of constraints. Although any unit can be selected as slack, it is preferable to select the unit with the largest capacity in order to have better chance to satisfy this constraint in one pass. Neglecting transmission losses P_L and considering P_1 as a slack unit, the following equation can be used:

$$P_1 = P_D - \sum_{i=2}^{n} P_i \tag{11}$$

When the term P_L is taken into account, Eq. (11) is converted to:

$$P_1 - \sum_{i=1}^{n} P_1 B_{1i} P_i - \sum_{i=2}^{n} P_i B_{i1} P_1 = P_D + \sum_{i=2}^{n}\sum_{j=2}^{n} P_i B_{ij} P_j + \sum_{i=2}^{n} B_{0i} P_i + B_{00} - \sum_{i=2}^{n} P_i \tag{12}$$

With extracting P_1 from the first summation, Eq. (12) becomes a quadratic equation as follows:

$$aP_1^2 + bP_1 + c = 0 \tag{13}$$

The analytical solution to this 2nd order polynomial equation can be obtained by finding the positive roots of the following general formula[2]:

$$P_1 = \frac{-b \pm \sqrt{b^2 - 4ac}}{2a}, \text{ and } b^2 - 4ac \geqslant 0 \tag{14}$$

where;

$$a = -B_{11} \tag{15}$$

$$b = 1 - \sum_{i=2}^{n} B_{1i} P_i - \sum_{i=2}^{n} P_i B_{i1} - B_{01} \tag{16}$$

$$c = \sum_{i=2}^{n} P_i - P_D - \sum_{i=2}^{n} \sum_{j=2}^{n} P_i B_{ij} P_j - \sum_{i=2}^{n} B_{0i} P_i - B_{00} \tag{17}$$

The slack generator approach is adopted in this study.

2.3 Generator Ramp Rate Limit Constraint

The conventional solution of the ELD problem does not take into account that the power output adjustment of the unit cannot be done instantaneously. Thus, for a more convenient solution, the ramp rate limit has to be considered, where the increasing and decreasing actions should happen within some specific ranges as follows:

$$P_i^{now} - P_i^{new} \leqslant R_i^{down} \tag{18}$$

$$P_i^{new} - P_i^{now} \leqslant R_i^{up} \tag{19}$$

where P_i^{now} and P_i^{new} are the existing and new power output of the ith generator, respectively. Also, R_i^{down} and R_i^{up} are the downward and upward ramp rate limits respectively [11]. These two equations can be included within Eq. (4) as follows [25]:

$$\max\left(P_i^{min}, P_i^{now} - R_i^{down}\right) \leqslant P_i^{new} \leqslant \min\left(P_i^{max}, P_i^{now} + R_i^{up}\right) \tag{20}$$

2.4 Prohibited Operating Zone Constraint

The prohibited operating zone is a phenomenon that may happen due to unit physical limitation, vibration in shaft bearing, steam valve opening, etc. [11], and causes some

---

[2]It is important to note that the solution of this analytical technique becomes valid only if P_1 passes some restrictions, like: located between lower and upper limits, positive and real, satisfies prohibited operating zones, satisfies downward and upward ramp rates, etc.

discontinuities on the fuel-cost curve. Thus, with this constraint, Eq. (4) becomes:

$$P_i^{min} \leqslant P_i \leqslant P_{i,j}^L$$
$$P_{i,j}^U \leqslant P_i \leqslant P_{i,j+1}^L$$
$$P_{i,\varkappa_i}^U \leqslant P_i \leqslant P_i^{max} \tag{21}$$

where $P_{i,j}^L$ and $P_{i,j}^U$ are respectively the lower and upper limits of the jth prohibited operating zone on the fuel-cost curve of the ith generating unit, and $\varkappa_i$ stands for the total number of the prohibited operating zones that are associated with the ith unit.

2.5 Emission Rates Constraint

Based on rules followed in each jurisdiction, there is an environmental regulation that requires power plants to not exceed the maximum allowable limits of the emission rates (*like: NOx, SOx, CO/CO2, UHC, etc.*) [26]. This constraint can be expressed as follows [8, 11]:

$$E_j(P_T) \leqslant MAL_j \tag{22}$$

where E_j and MAL_j stand for the emission rate and the maximum allowable limit of the jth gas, respectively.

In addition to the above constraints, the objective function could be subject to many other constraints based on the type and operational philosophy of the electrical power plant operators. Some of these constraints are: spinning reserve constraint, line flow constraint, reservoir storage limits, water balance equation, network security constraint, etc. [5, 8, 11, 14].

Moreover, optimizing this objective function requires using sub-algorithm to deal with all of these constraints in order to get a feasible solution. For side constraints, the independent or decision variables $\{P_1, P_2, \ldots, P_i, \ldots, P_n\}$ can be easily satisfied by bounding the randomly generated values between the lower and upper limits,[3] as shown in Eq. (4). The other constraints (i.e., the equality and inequality constraints) can be satisfied by using different constraint-handling techniques [27–29].

3 Hybridizing BBO, SA and SQP Algorithms

The core of this hybridization is to have a good balance between the exploration and exploitation levels. In this proposed algorithm, two phases are suggested. Firstly, modifying the migration stage of BBO by hybridizing it with the Metropolis criterion of SA, so that the bad migrated islands can be checked whether they are worth being

---

[3]This statement becomes invalid if ramp rate limits, prohibited operating zones and transmission losses P_L are considered.

candidate solutions or not. Secondly, the best obtained solutions are fine-tuned by recycling them through SQP. If the tuned elites are improved, then they will take the positions of the worst individuals in order to keep the original elites too. The final name of this hybrid algorithm, which is a combination of three different algorithms, is MpBBO-SQP. It has been tried to collect the main strengths of each algorithm into one algorithm; and, at the same time, the main weaknesses are rejected. However, one of the inherent problems that are faced with most hybrid algorithms is that the processing times are relatively higher than that of the classical algorithms.

In order to fully understand the given subject, it is important to describe first the general mechanisms of both BBO and SA algorithms[4].

3.1 Simulated Annealing Algorithm

In metallurgy and materials science, the word "annealing" means a heat process that controls the properties of the metal (*like: ductility, strength, hardness, etc.*) by heating it up to a specific temperature (*above recrystallization temperature*), maintaining that temperature for a certain period, and then allowed to cool slowly. By this process, crystals will be formed in good shape with the lowest internal energy, and hence the metal will settle on a crystalline state. In case the cooling rate is very fast, the metal will be on a polycrystalline state, so that the high internal energy will deform the crystals' structure [29].

This slow cooling process inspired Kirkpatrick et al. [31] to propose their novel single-point global optimization algorithm, which was presented in 1983 as the "simulated annealing (SA) algorithm". In addition, independently the method was presented in 1985 by Vlado Černý [32]. The core of this derivative free nature-inspired optimization algorithm is the statistical mechanics that was demonstrated by Metropolis et al. in 1953 [33] using the concept of Boltzmann's probability distribution. It states that if a system is maintained in a thermal equilibrium at temperature $\check{T}$, the probabilistic distribution $\check{P}$ of its energy $\check{E}$ can be achieved by [29]:

$$\check{P}(\check{E}) = e^{\frac{-\Delta \check{E}}{k_B \check{T}}} \qquad (23)$$

where k_B is a Boltzmann's constant; for simplicity, it is set to one. $\Delta \check{E}$ is the difference in energy, which is translated in SA as the difference in cost function. It can be calculated as:

$$\Delta \check{E} = f(X^{new}) - f(X^{old}) \qquad (24)$$

---

[4]SQP is skipped because it is very popular and detailed explanation can be easily found in many mathematical optimization books; like [30].

For minimization problems, the new design point X^{new} is directly accepted if the following condition is satisfied:

$$f(X^{new}) \leqslant f(X^{old}) \tag{25}$$

In case Eq. (25) is not satisfied, then the new design point X^{new} will not be directly rejected. Rather, It has another chance to be accepted if it passes the Metropolis criterion.

From Eq. (23), the probability to accept X^{new} will increase as the molten metal is heated to a very high temperature $\breve{T}$, and that chance gradually decreases during annealing or slow cooling process. This process will avoid trapping into local optima when $\breve{T}$ is high.

Based on that fact, it is very important to initialize SA with high $\breve{T}$. But the problem is in reality here: "how much?". Actually, it has been found that the good initial temperature $\breve{T}_o$ for some objective functions could be unsuitable for other objective functions. Thus, the determination of $\breve{T}_o$ itself becomes a challenge. To solve this issue, many researchers estimate $\breve{T}_o$ based on the average cost of different randomly generated individuals; *as will be discussed later.*

The conventional SA algorithm is constructed with two searching loops. The external loops are coded as a number of cycles or stages z, while the internal loops are set as a number of iterations l. That is, SA is initialized with high $\breve{T}_o$ for the first cycle, and decreases with specific cooling rate.[5] During each cycle, the fitness is enhanced with certain number of iterations. Therefore, convergence speed and solution quality highly depend on the setting of $\breve{T}_o$ and how it is cooled down. Moreover, the maximum limits of the external and internal loops (*i.e., the number of cycles z and iterations l*) are also important settings in SA. Small z and l lead to fast computation but with low performance. In contrast, large z and l improve its performance but at the cost of a huge amount of CPU time. Thus, the process is like a compromise between the solution quality and processing speed. The CPU time can be reduced if these z and l loops are terminated once the solution tolerance is satisfied ($|\Delta \breve{E}| \leqslant \varepsilon$). However, this option should be turned-off in order to have a fair comparison with other algorithms when the processing speed is also considered as a one of the performance criteria.

3.2 Biogeography-Based Optimization Algorithm

The mechanism of this new population-based evolutionary algorithm is inspired by an old scientific study in biogeography[6] that was conducted by the ecologists Robert

---

[5]In the literature, different cooling strategies are proposed for this purpose.

[6]Biogeography is a branch of biology science, and it is a synthetic discipline, relying heavily on theory and data from ecology, population biology, systematics, evolutionary biology, and the earth sciences [34]. Biogeography seeks to describe, analyze and explain the geographic patterns

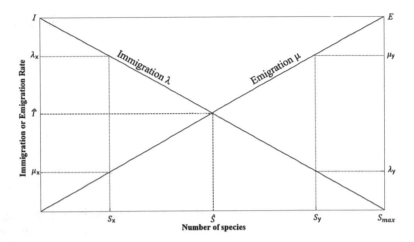

Fig. 1 Simplified equilibrium model of a biota of a single island

H. MacArthur and Edward O. Wilson between 1960–1967 [37, 38]. This study is known as "The Theory of Island Biogeography", and they proposed that the dynamic equilibrium between immigrated and extinct species controls the endemic species on the isolated islands.[7]

The immigration rate λ and the emigration rate μ can be set in many ways.[8] In order to simplify the mathematical process, MacArthur and Wilson used a simplified linear migration model with similar maximum immigration and emigration rates (i.e., $I = \lambda_{max} = E = \mu_{max}$) as shown in Fig. 1. $\hat{T}$ is the species turnover rate, which happens when the species density settles on the equilibrium state $\hat{S}$, where S_{max} is the maximum number of endemic species on that island [37, 38, 42]. λ_{max} or I happens when there is no available species on the ith island, and μ_{max} or E happens when all the nests are occupied by the emigrated species from the mainland and/or other islands [44].

In BBO, the population size k is simplified to be equal to the maximum number of species S_{max}. Therefore, λ_i and μ_i of the basic migration model, given in Fig. 1, can be determined as follows:

---

and changing distributions of ecosystems and fossil species of plants (flora) and animals (fauna) through geological space and time [35, 36].

[7]In island biogeography, the word "island" could be aquatic island, desert oasis, individual plants, caves, lakes or ponds, mountain-tops (sky-islands), microcosms or even patches of terrestrial ecosystems [39, 40].

[8]The emigration and immigration rates can be modeled as exponential, logistic, linear, etc. [34, 41, 42]. Also, the maximum emigration and immigration rates can be unequal (i.e., $I \neq E$) [38, 42]. Moreover, the equilibrium location $\hat{S}$ can be shifted to the right or left side based on the type of rate functions, the area of island and/or the distance or isolation between the recipient island and the source island or mainland [34, 38, 43].

$$\mu_i = \left(\frac{E}{k}\right) i \tag{26}$$

$$\lambda_i = 1 - \mu_i = I\left(1 - \frac{i}{k}\right) \tag{27}$$

Suppose at time t the island contains i species with probability $Pr_i(t)$, then the variation of the probability from t to $(t + \Delta t)$ can be described as follows [37, 38]:

$$Pr_i(t + \Delta t) = Pr_i(t)(1 - \lambda_i \Delta t - \mu_i \Delta t) + Pr_{i-1}(t)\lambda_{i-1}\Delta t + Pr_{i+1}(t)\mu_{i+1}\Delta t \tag{28}$$

Considering Eq. (28), to have i species at time $(t + \Delta t)$, one of the following three conditions should be satisfied [37, 38]:

1. i species at time t, and no migrated species during the interval Δt;
2. $(i - 1)$ species at time t, and one species immigrated;
3. $(i + 1)$ species at time t, and one species emigrated.

From calculus, it is known that the ratio $\left(\frac{\Delta Pr_i}{\Delta t}\right)$ approaches $\dot{Pr}_i(t)$ as $\Delta t \to 0$:

$$\dot{Pr}_i(t) \cong \lim_{\Delta t \to 0} \frac{Pr_i(t + \Delta t) - Pr_i(t)}{\Delta t}$$
$$\cong -(\lambda_i + \mu_i)Pr_i(t) + \lambda_{i-1}Pr_{i-1}(t) + \mu_{i+1}Pr_{i+1}(t) \tag{29}$$

By considering the preceding three conditions, Eq. (29) can be re-expressed with three cases as follows:

$$\dot{Pr}_i(t) = \begin{cases} -(\lambda_i + \mu_i)Pr_i(t) + \mu_{i+1}Pr_{i+1}(t), & \text{if } i = 0 \\ -(\lambda_i + \mu_i)Pr_i(t) + \lambda_{i-1}Pr_{i-1}(t) + \mu_{i+1}Pr_{i+1}(t), & \text{if } 1 \leqslant i \leqslant k - 1 \\ -(\lambda_i + \mu_i)Pr_i(t) + \lambda_{i-1}Pr_{i-1}(t), & \text{if } i = k \end{cases} \tag{30}$$

The value of $\dot{Pr}_i(t)$ can also be determined by using the matrix technique in [18], which was successfully proved in [45]. Thus, using the known values of $Pr_i(t)$ and $\dot{Pr}_i(t)$, the value of $Pr_i(t + \Delta t)$ given in Eq. (28) can be approximated as follows:

$$Pr_i(t + \Delta t) \cong Pr_i(t) + \dot{Pr}_i(t)\Delta t \tag{31}$$

Equation (31) is the final form that should be used in the BBO program. To find $\dot{Pr}_i(t)$, two methods have been used by Simon in [18]: the first one is by solving Eq. (30) numerically while the other one can be directly applied through the following theorem:

Theorem 1 *The steady-state value for the probability of the number of each species is given by:*

$$Pr(\infty) = \frac{v}{\sum\limits_{w=1}^{k+1} v_w} \tag{32}$$

The eigenvector v can be computed as follows:

$$v = [v_1, v_2, \ldots, v_{k+1}]^T, \quad T \text{ means transpose}$$

$$v_w = \begin{cases} \frac{k!}{(k+1-w)!(w-1)!}, & \text{for } w = 1, 2, \ldots, w' \\ v_{k+2-w}, & \text{for } w = w'+1, \ldots, k+1 \end{cases}$$

$$\text{where: } w' = \left\lceil \frac{k+1}{2} \right\rceil \tag{33}$$

Although the second method is easier and $Pr_i(t)$ can be directly computed without any iterations, this approach is not preferable in most popular programs, such as MATLAB, Python, C/C++, Octave, etc., because $k = \infty$ when $k > 170$. This infinity problem can be solved if an additional sub-algorithm is used. However, dealing with long product operations will require extra CPU time [23]. Based on that, the numerical method is flexible and more convenient, and hence adopted in this study.

In BBO, the objective function can be optimized if each island is considered as one individual, while the independent variables of each individual are dealt as features. The solutions can be enhanced if these features are distributed between the source and recipient islands. The source island could become recipient island for other better islands [18]. That is, the richness of species on an island is decided through a probabilistic process. If many good biotic and abiotic features[9] are available on an island, then it will be a good land for immigrants. Each feature is called suitability index variable (*SIV*), and represents an independent variable of such a problem in BBO. The island suitability index (*ISI*) is the dependent variable that varies as changing the elements of the *SIV* vector. Because BBO is a population-based algorithm, so optimizing n-dimensional problem with k-individuals can be mathematically represented as follows:

$$ISI_i = f(SIV_{i1}, SIV_{i2}, \ldots, SIV_{in}), \quad i = 1, 2, \ldots, k \tag{34}$$

Once the initialization stage is completed, the BBO algorithm should pass some sub-algorithms.

---

[9]Biotic factors: predation, competition, interactions, etc. Abiotic factors: wind, water, sunlight, temperature, pressure, soil, etc. [46].

3.2.1 Migration Stage

The main idea of this stage is to share the good features of the rich islands in order to modify the poor islands. Because the selection is done through a probabilistic process, so the ith island is likely to be selected as a source of modification if ISI_i is high, and vice versa for the jth recipient island. From Fig. 1, low λ_i and high μ_i mean a large number of endemic species are available on the ith island. Thus, the solution ISI_i is high. As an example, the point S_x is located before $\hat{S}$, so λ_x is high, μ_x is low and ISI_x is considered as a poor solution; while the point S_y is located after $\hat{S}$, so λ_y is low, μ_y is high and ISI_y is considered as a good solution. Based on that, μ_i and λ_i are used as indications of quality of the solution.

Through the migration process, the low ISI islands could be improved per each new generation, and at the same time the quality of the solution of the best islands are kept away from any corruption.

The original BBO algorithm comes with four migration forms, as described in [47, 48], and called partial, simplified partial, single, and simplified single migration based (PMB, SPMB, SMB, and SSMB) models. The first published BBO paper used the PMB model [18], which is graphically illustrated in Fig. 2. As can be clearly seen from this figure, these rich and poor islands act as source and recipient of those migrated n SIV. Each $SIV s$ of poor islands is updated by $SIV\sigma$ that is probabilistically selected from one rich island. For the SPMB model, the n SIV of poor islands are updated from the first best island(s), which in turn increases the probability to trap into local optimums. The migration process of the SMB and SSMB models are respectively similar to the PMB and SPMB models with one main difference: only one randomly selected $SIV s$ of each poor island is modified. The last two models are faster, but with low solution quality. This study will consider all the essential modifications presented in [23] as a basis to the proposed MpBBO-SQP algorithm. Thus, the raw BBO algorithm (before be hybridized with SA and SQP) can save around 32.32 % of its total CPU time and with better performance than that of PMB-

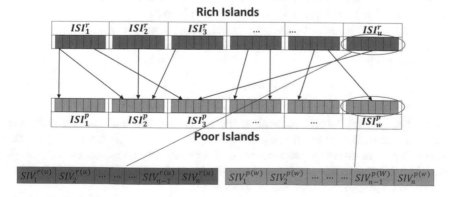

Fig. 2 Migration process between different islands in BBO

BBO model. The migration process, *which is used in this study*, is described by Algorithm 1.

Algorithm 1 Partial Migration Pseudocode

Require: Let ISI_i denote the ith population member and contains n features
Require: Define emigration rate μ_i and immigration rate λ_i for each member
  1: **for** $i \leftarrow 1$ *to* k **do** {where k = number of islands or individuals, see Eq. (34)}
  2:    **for** $s \leftarrow 1$ *to* n **do** {where n = number of features "*SIV*" or design variables}
  3:       Use λ_i to probabilistically select the immigrating island ISI_i
  4:       **for** $j \leftarrow 1$ *to* k **do** {Break once ISI_j is selected}
  5:         Use μ_j to probabilistically decide whether to emigrate to ISI_i
  6:         **if** ISI_j is selected **then** {where $ISI_i \neq ISI_j$}
  7:           Randomly select an *SIV* σ from ISI_j
  8:           Replace a random *SIV* s in ISI_i with *SIV* σ
  9:         **end if**
10:       **end for**
11:       next *SIV*
12:    **end for**
13:    next island
14: **end for**

3.2.2 Mutation Stage

As with many nature-inspired algorithms, this stage is very essential to increase the exploration level. The mutation process of the BBO algorithm can be defined as some random natural events that affect the availability of the biotic and abiotic features on an isolated island, which in turn reflects on the total endemic species on that island. These events could be positive (*like: shipwreck and wind pollination*) that increase the species density, or could be negative (*like: volcanoes, diseases and earthquakes*).

In BBO, the species count probabilities Pr_i is used exclusively to find the mutation rate m_i [18]. Thus, there are many choices available to the researchers to select their preferable mutation rate, like using Gaussian, Cauchy and Lèvy mutation operators as in [21]. The original mutation rate, *which is also used in this study*, is described as follows [18]:

$$m_i = m_{max}\left(1 - \frac{Pr_i}{Pr_{max}}\right) \tag{35}$$

where Pr_{max} is the largest element of the Pr vector, and m_{max} is a user-defined maximum allowable value that m_i can reach. As can be seen from Eq. (35), the mutation rate is inversely proportional to the probability rate (i.e., $m_i \propto^{-1} Pr_i$), where m_i is equal to m_{max} at ($Pr_i = 0$), and it is equal to 0 at the largest element of Pr. It can be graphically represented as shown in Fig. 3.

The mutation rate will flip the bell-shape graph of the probability rate. The main objective of using m_i rather than Pr_i is to have better control on the targeted islands

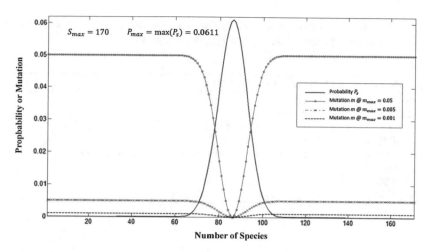

Fig. 3 Comparison between Pr and m at different m_{max}

for the mutation stage. That is, the islands located in or near the equilibrium point $\hat{S}$ will be preserved, while the other islands sorted on both sides will have higher chance to be mutated, and hence could be improved. The mutation process is described by Algorithm 2.

Algorithm 2 Original Mutation Pseudocode

1: **for** $i \leftarrow 1$ *to* k **do** {where k = number of islands or individuals, see Eq. (34)}
2:   Calculate probability Pr_i based on λ_i and μ_i {by iterative or eigenvector method}
3:   Calculate mutation rate m_i {using Eq. (35)}
4:   **if** rand $< m_i$ and $i \geqslant R_m$ **then** {R_m is a user defined mutation range}
5:     Replace n SIV vector of ISI_i with a randomly generated n SIV vector
6:   **end if**
7: **end for**

3.2.3 Clear Duplication Stage

If this optional stage is used in BBO, then the features' diversity increase. This is because the emigrated *SIV* will take the same value and place in other island(s), so they may become inactive. In this situation, the exploration level will decrease and the algorithm may quickly settle on a non-global optimum solution.[10] The main purpose of this stage is to check all n *SIV* of all k *ISI* whether they are duplicated or

---

[10]The blended BBO, given in [49], has the ability to avoid this duplication phenomenon.

not. If any duplicated feature is indicated, then it will be replaced by a new generated random value. This process is described by Algorithm 3 [50].

Algorithm 3 Clear Duplication Pseudocode

Require: Check all n *SIV* on all k *ISI*
 1: **while** there is a duplicated *SIV* **do**
 2:    **for** $i \leftarrow 1$ *to* k **do** {where k = number of islands or individuals, see Eq. (34)}
 3:       **if** any duplicated *SIV s* is detected **then**
 4:          Replace the duplicated *SIV s* in *ISI$_i$* with a randomly generated *SIV σ*
 5:       **end if**
 6:    **end for**
 7: **end while**

3.2.4   Elitism Stage

Suppose that the good individuals of the gth generation have been ruined by the previous BBO stages. Then, they will be lost forever if this optional stage is not activated in the BBO program. This stage will provide a rollback feature to rescue the last state of the corrupted best islands (elites) and recycled again for the next generation [51].

3.3   Hybrid MpBBO-SQP Algorithm

In the classical BBO algorithm, the emigrated SIV_σ from ISI_j to ISI_i are directly confirmed. Although the elitism stage can recover the elites before being corrupted, the ruined features of the non-elite individuals will be lost forever. To have control on the emigrated features, the migration stage is modified by adding the Metropolis criterion of SA as a checking unit. If $ISI_i(SIV^{new}) \leqslant ISI_i(SIV^{old})$, the ith individual will be directly updated with the new decision vector. Otherwise, it will be updated only if it passes the Metropolis criterion. Additionally, this given chance will gradually decrease as the generations number increases, because the temperature T_g will decrease and hence satisfying Eq. (23) becomes hard. This is why the cooling rate is considered as a one of the most important settings of SA [29].

3.3.1   Cooling Strategies

As stated earlier, the good initial temperature is not the same for all optimization problems. A specific value could be good for some arbitrary/real problems, but it may become very bad for other problems. The general formula given in [29] is used here to estimate the good $\check{T}_o$ based on the average of four randomly generated individuals:

$$\check{T}_o = \frac{\text{sum}\,(ISI_1 \rightarrow ISI_4)}{4} \tag{36}$$

where $ISI_1 \rightarrow ISI_4$ are the 1st four best solutions of the initialization stage of the BBO sub-algorithm.

The scheduling simplicity, solution quality, speed, etc., are very important factors that affect on the selection criteria. Some of the most popular types are briefly described below [27, 29, 52]:

A. Linear Cooling Schedule

It is very simple type. The gth cooling temperature $\check{T}_g$ is calculated as follows:

$$\check{T}(g) = \max\left(\check{T}_o - \xi \times g, \check{T}_{min}\right) \tag{37}$$

where $\check{T}_{min}$ is the smallest acceptable value that should not be exceeded by $\check{T}_g$; it is set to 0.00001. ξ is the slope of the above linear equation, and it can be determined as follows:

$$\xi = \frac{\check{T}_o}{G} \tag{38}$$

where G is the total number of generations initialized for the BBO sub-algorithm.

B. Exponential Cooling Schedule

In general, this type can effectively achieve a compromise between the simplicity, solution quality and speed. This is why it becomes very popular in the literature. It can be calculated as follows:

$$\check{T}(g) = r \times \check{T}(g-1) \tag{39}$$

where $\check{T}(g-1)$ is equal to $\check{T}_o$ when $(g = 1)$. Based on the extracted results and the recommendations given in [27, 29], the temperature reduction factor r is set to 0.6.

C. Inverse Cooling Schedule

The following inverse function used the past temperature $\check{T}(g-1)$ to calculate the current temperature $\check{T}(g)$ [53]:

$$\check{T}(g) = \frac{\check{T}(g-1)}{[1 + \varsigma \times \check{T}(g-1)]} \tag{40}$$

where ς is a small constant, and it is set to 0.005 in this study [27].

D. Inverse Linear Cooling Schedule

The main goal of designing this type of schedules is to have a fast simulated annealing (FSA) algorithm [54]. It can be easily coded as follows:

Fig. 4 Five different cooling
strategies of the simulated
annealing (SA) algorithm

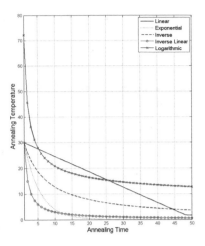

$$\check{T}(g) = \frac{\check{T}_o}{g} \qquad (41)$$

E. Logarithmic Cooling Schedule

This logarithmic function is mathematically expressed as follows [55]:

$$\check{T}(g) = \frac{\rho}{\ln(g + d)} \qquad (42)$$

where d is a constant, and usually set equal to one. Although ρ can also be coded as
a constant, it is selected as the biggest individual of the initialization stage [52]:

$$\rho = \max(ISI_1 \rightarrow ISI_k) \qquad (43)$$

Figure 4 graphically summarizes the previous five cooling strategies.

3.3.2 Fine Tuning with SQP Phase

This idea has been adopted in some optimization algorithms. Based on previous
conducted studies, it has been found that the fine tuning stage, by linear/non-linear
programming (NP/NLP) algorithms, can effectively enhance the solutions quality,
and the hybrid algorithms will require less population size k and generations G to
settle on the global or near-global solution [56–58].

Figure 5 shows how the SQP sub-algorithm acts within MpBBO-SQP. The elites
of each generation will be recycled through SQP. The original elites (*i.e., obtained by
MpBBO phase*) will not be overwritten after having been enhanced by SQP. Instead,
the worst individuals will be replaced by these fine-tuned elites. The performance
of SQP highly depends on the parameter *elit*, and the processing time of SQP will
increase as *elit* increases.

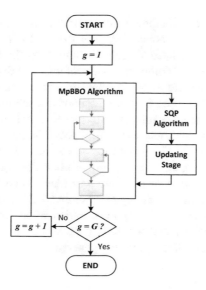

Fig. 5 Flowchart of the proposed operation philosophy of the SQP phase within MpBBO-SQP. This phase can be activated at the beginning of the algorithm (i.e., *when g* = 1), or after completing a specific number of generations (*in order to save CPU time*)

3.3.3 Final Structure of MpBBO-SQP

Both SA and BBO are probabilistic meta-heuristic optimizers, but each algorithm has its own strengths and weaknesses. SA is a single-point searcher while BBO is a population-based searcher. SA could become faster if it is initialized with the same number of generations, but BBO is more capable to explore the search space. However, the exploitation level of BBO is higher than the exploration level [19–22]. This is because of the inherent limitations of its migration process [23]. This unbalanced factor can be treated by applying the main feature of SA, where at the first cycles of the annealing process the temperature $\check{T}$ are high and thus the probability to discover other parts of the search space is high. Although SA can be markedly affected by its parameters, like initial temperature $\check{T}_o$ and its associated cooling rate, BBO has more immunity against its own parameters.

Because SA and BBO are stochastic-based optimization algorithms, they are well known as time-consuming techniques. On the other hand, conventional algorithms are gradient-based techniques, but they are highly prone to be trapped into local optimums if they are not initialized with good starting points. Thus, in order to accelerate the convergence speed and accuracy, the hybridization between traditional and non-traditional algorithms is currently one of the popular approaches [56–60].

In general, hybridization approaches could increase the total amount of the consumed CPU time. However, MpBBO-SQP is relatively very fast algorithm, because the internal searching loops l of the SA sub-algorithm are excluded and compensated by receiving multiple design points from the BBO sub-algorithm without getting affecting the overall performance of MpBBO-SQP. Also, the essential modifications considered in [23] can save additional amount of the CPU time.

Algorithm 4 describes the whole mechanism of the proposed hybrid MpBBO-SQP algorithm. As can be seen from the pseudocode, the migration stage of BBO is extended to include the Metropolis criterion of SA. Thus, the migrated n SIV between k ISI will not be confirmed unless they pass this test. To accomplish this, two temporary matrices (M_1 and M_2 → with size $k \times n$) are used to store the population before and after completing the migration process of the gth generation. The ith place of the population is filled with the modified individual stored in M_2 if it shows better quality than its old state (*i.e. replacing ISI_i^{old} stored in M_1 with ISI_i^{new} stored in M_2*) or if it successfully passes the Metropolis criterion when $ISI_i^{new} > ISI_i^{old}$. Once the migration stage is completed and checked by the Metropolis criterion, the elites are fine-tuned by SQP before starting the next generation.

4 Solving ELD Problems

The performance of BBO with and without hybridization is evaluated using three different test systems. The algorithms' parameters for each system are listed in Table 1. All three test systems have been modelled using the quadratic cost function and considering the valve-point loading effects, see Eq. (3), where the required data and information about these systems are available in [16].

All simulations are carried out on the same machine with the following specifications: MATLAB R2011a using Intel Pentium E5300—2.60 GHz and 4 GB RAM with 32-bit Windows XP SP3 operating system.

Table 1 The algorithms parameters for each test case

Initialization settings[a]	3-units test system		13-units test system		40-units test system	
	Without SQP	With SQP	Without SQP	With SQP	Without SQP	With SQP
Population size (k)	20	5	20	15	60	20
Iterations no. (G)	50	20	250	40	1000	500
Total trials (Tr)	50	50	50	50	50	50
Mutation (m_{max}) [13]	0.007	0.007	0.009	0.009	0.007	0.007
Elitism (*elit*)	1	1	1	4	3	8

[a]The initial population is randomly generated for all the test systems

---

Algorithm 4 Hybrid MpBBO-SQP Algorithm Pseudocode

Require: Initialization stage with all the parameters
1: Find λ_i and μ_i rates, then Pr_i and m_i rates
2: Generate k islands with non-duplicated n *SIV*
3: Sort and map the population to the species count (i.e., $ISI_1 = ISI_{best}$ is coupled with μ_{max} or λ_{min}, and continue till reaching $ISI_k = ISI_{worst}$)
4: **for** $g \leftarrow 1$ *to* G **do** {where $G =$ number of generations}
5:    **if** $g = 1$ **then**
6:        Find the initial temperature $\breve{T}_o$ based on the average of the 1st four best solutions (using Eq. (36))
7:    **else**
8:        Updated the temperature for the gth generation
9:    **end if**
10:    Save the required best solutions "elites" to be recycled again in the next generation
11:    Save the vectors of all the individuals (before migration) in a temporary matrix M_1 with size $k \times n$ and their cost functions in a temporary vector V_1 with length k
12:    Do migration (refer to Algorithm 1)
13:    Save the vectors of all the individuals (after migration) in a temporary matrix M_2 with size $k \times n$ and their cost functions in a temporary vector V_2 with length k
14:    **for** $i \leftarrow 1$ *to* k **do** {where $k =$ number of islands or individuals}
15:        Calculate $\Delta\breve{E}_i = V_2(i) - V_1(i)$
16:        **if** $\Delta\breve{E}_i > 0$ **then**
17:            Apply the Metropolis criterion $\breve{P}(\breve{E}) = e^{\frac{-\Delta\breve{E}_i}{k_B\breve{T}(g)}}$
18:            **if** $\breve{P}(\breve{E}) >$ rand **then**
19:                Accept $M_2(i, 1 \rightarrow n)$ vector of matrix M_2 as an updated individual for ISI_i
20:            **else**
21:                Re-select the past $M_1(i, 1 \rightarrow n)$ vector of matrix M_1 as a confirmed individual for ISI_i
22:            **end if**
23:        **end if**
24:    **end for**
25:    Update the population with sorting and mapping
26:    Select the best individuals (*elit*)
27:    **for** $j \leftarrow 1$ *to* *elit* **do** {where *elit* = the best individuals}
28:        Tune the jth elite by SQP
29:        **if** $ISI_j^{tuned} \leqslant ISI_j^{untuned}$ **then**
30:            Insert ISI_j^{tuned} in the population by taking the place of the worst individuals
31:        **else**
32:            Neglect ISI_j^{tuned}
33:        **end if**
34:    **end for**
35:    Update the population with sorting and mapping
36:    Do mutation (refer to Algorithms 2)
37:    Clear any duplicated *SIV* (refer to Algorithm 3)
38:    Update the population with sorting and mapping
39:    **if** $g > 1$ **then**
40:        Replace the worst *ISI* with the previous good *ISI* that are saved in the elitism stage
41:        Update the population with sorting and mapping
42:    **end if**
43: **end for**
44: Display the best individual

Table 2 Comparison between BBO algorithms for test case I

Generating unit	BBO-EM[a]	MpBBO					MpBBO-SQP
	PMB	Lin.	Exp.	Inv.	Inv. Lin.	Log.	Exp.
P_1 (MW)	300.155	300.084	300.375	299.632	300.331	300.302	300.267
P_2 (MW)	149.925	149.982	149.802	150.398	149.780	149.783	149.733
P_3 (MW)	399.920	399.934	399.823	399.970	399.889	399.915	400.000
$\sum P_i$ (MW)	850.000	850.000	850.000	850.000	850.000	850.000	850.000
Best cost ($/h)	8234.22	8234.24	8234.20	8234.46	8234.15	8234.14	**8234.07**
Mean ($/h)	8252.78	8254.33	**8249.37**	8250.34	8254.91	8254.43	8260.67
Median ($/h)	8248.36	8252.37	8243.98	8246.37	8251.09	8249.23	**8241.59**
Std. Dev. ($/h)	17.5654	**15.0395**	15.6774	15.8421	16.3213	17.6732	45.2105
Avg. CPU time (s)	**0.05233**	0.05976	0.06033	0.06143	0.05997	0.05982	0.26762

[a]The acronym BBO-EM stands for BBO with essential modifications. This algorithm has been previously used in [23] and won against the original BBO algorithm. It is used here just to show the superiority of the proposed MpBBO-SQP algorithm

4.1 Test Case I—3 Generating Units

This system contains three generating units with the load demand of 850 MW. Table 2 shows the results obtained by different BBO versions. Based on the mean of MpBBO, the exponential cooling rate is selected for MpBBO-SQP. It can be clearly seen from this table that MpBBO-SQP can converge to the best solution, but it consumes more CPU time as compared with BBO-EM and MpBBO. Also, it can be observed that due to few number of generations MpBBO-SQP may be trapped into local optima. However, this happens just for some few trials as can be concluded from the median. This can be easily avoided by increasing G and/or k. Figure 6 shows the fitness curves of this test system, which proved again the superiority of the MpBBO-SQP approach.

Table 3 shows an extended comparison with other optimization algorithms presented in the literature. As can be obviously seen from this table, MpBBO-SQP can reach the best known solution with less k and G.

4.2 Test Case II—13 Generating Units

This is the second system, which is relatively harder than the first system. The load demand that has to be satisfied is 1800 MW. Table 4 shows the results obtained by BBO-EM, MpBBO and MpBBO-SQP after 50 trials. Again, MpBBO-SQP is executed with the exponential cooling rate, because it shows better mean as compared with the other four schedules. With fair CPU time comparison, it has been found that MpBBO-SQP consumes more CPU time. However, it significantly outperforms the other two algorithms; as can be seen in Fig. 7. This superiority can also be observed

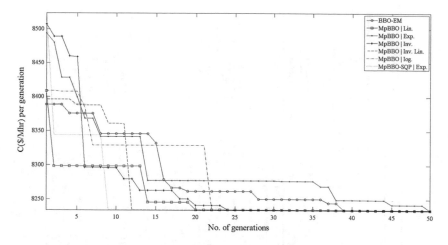

Fig. 6 The fitness curves of BBO versions for test case I

Table 3 Comparison between MpBBO-SQP and other algorithms for test case I

Algorithm type	Performance criteria		
	Best cost ($/h)	Population size (k)	Iterations no. (G)
EP [61]	**8234.07**	30	50
GAB [16]	8234.08	20	50
GAF [16]	**8234.07**	20	50
CEP [16]	**8234.07**	20	50
FEP [16]	**8234.07**	20	50
MFEP [16]	8234.08	20	50
IFEP [16]	**8234.07**	20	50
MPSO [62]	**8234.07**	20	150
GA [13]	8239.20	300	150
PSO [13]	8234.72	300	150
BBO [13]	8234.08	300	150
MpBBO-SQP	**8234.07**	5	**20**

in Table 5. Among these 17 different competitive optimization techniques, MpBBO-SQP can reach 17963.8 $/h with very small k and G.

4.3 Test Case III—40 Generating Units

With 40 generating units, this system is considered as a one of the largest test cases available in the literature. The load demand of this ELD problem is 10500 MW.

Table 4 Comparison between BBO algorithms for test case II

Generating unit	BBO-EM[a]	MpBBO					MpBBO-SQP
	PMB	Lin.	Exp.	Inv.	Inv. Lin.	Log.	Exp.
P_1 (MW)	449.268	449.099	539.574	538.214	538.804	449.194	628.319
P_2 (MW)	305.198	225.160	150.093	299.622	79.921	146.605	149.599
P_3 (MW)	145.659	221.228	228.265	70.523	148.985	149.103	222.751
P_4 (MW)	109.085	114.141	158.729	111.007	161.286	159.364	109.865
P_5 (MW)	111.583	109.271	60.143	62.090	107.561	161.413	109.866
P_6 (MW)	109.165	109.234	110.723	60.057	159.572	163.772	109.867
P_7 (MW)	66.301	158.037	110.242	110.634	159.795	108.754	109.867
P_8 (MW)	109.126	114.962	61.883	160.157	61.197	158.759	60.000
P_9 (MW)	159.715	62.576	109.950	155.804	109.449	60.633	109.867
P_{10} (MW)	78.143	80.973	42.789	40.405	82.760	42.496	40.000
P_{11} (MW)	44.689	42.393	79.005	44.654	40.547	83.474	40.000
P_{12} (MW)	56.509	55.606	56.841	91.481	58.033	55.353	55.000
P_{13} (MW)	55.557	57.322	91.762	55.352	92.089	61.081	55.000
$\sum P_i$ (MW)	1800.00	1800.00	1800.00	1800.00	1800.00	1800.00	1800.00
Best cost ($/h)	18227.5	18226.9	18153.9	18185.0	18226.6	18267.0	**17963.8**
Mean ($/h)	18393.1	18386.0	18301.6	18308.8	18325.4	18397.7	**18070.0**
Median ($/h)	18380.4	18389.8	18290.9	18304.3	18321.7	18402.8	**18073.3**
Std. Dev. ($/h)	81.0980	61.6227	74.5877	55.3060	61.9199	58.3879	**40.7717**
Avg. CPU time (s)	**0.57700**	0.62722	0.82779	0.83726	0.73686	0.63166	4.47230

[a]See the footnote of Table 2

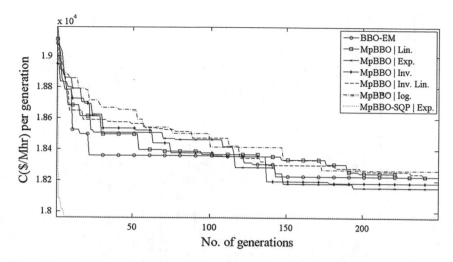

Fig. 7 The fitness curves of BBO versions for test case II

Table 5 Comparison between MpBBO-SQP and other algorithms for test case II

Algorithm type	Performance criteria		
	Best cost ($/h)	Population size (k)	Iterations no. (G)
CEP [16]	18048.2	30	800
FEP [16]	18018.0	30	800
MFEP [16]	18028.1	30	800
IFEP [16]	17994.1	30	800
PSO-SQP [56]	17969.9	100	100
CGA_MU [63]	17975.3	30	30 × 3000
IGA_MU [63]	17964.0	**5**	30 × 3000
PSO [64]	18014.2	20	250
PPSO [64]	17971.0	20	250
MPPSO [64]	17976.2	20	250
APPSO [64]	17978.9	20	250
DPSO [64]	17976.3	20	250
GAPSSQP [58]	17964.25	100	1000
ABC [12]	17963.9	300	200
FAPSO-NM [65]	**17963.8**	26	300
FAPSO-VDE [66]	**17963.8**	26	100
BBO [13]	17965.8	50	100
MpBBO-SQP	**17963.8**	15	**40**

Table 6 Comparison between BBO algorithms for test case III

Generating unit	BBO-EM[a]	MpBBO					MpBBO-SQP
	PMB	Lin.	Exp.	Inv.	Inv. Lin.	Log.	Inv. Lin.
P_1 (MW)	109.9829	107.1563	110.4708	113.5721	112.3417	110.6294	110.8006
P_2 (MW)	112.9480	105.3991	109.8269	110.6250	110.2898	111.5661	110.8009
P_3 (MW)	60.8319	101.0059	92.6172	118.7452	96.7447	98.6038	97.4002
P_4 (MW)	177.5861	179.2414	177.6211	183.7266	180.1699	180.5318	179.7331
P_5 (MW)	89.5334	92.0468	88.5719	89.1711	91.1842	96.0074	90.7899
P_6 (MW)	111.2320	109.1939	131.7801	138.4395	106.1829	135.3823	140.0000
P_7 (MW)	259.8986	256.5497	292.8111	275.7522	257.3219	280.3666	259.6003
P_8 (MW)	283.0137	283.1697	284.6723	291.4189	285.2167	285.0032	284.6004
P_9 (MW)	283.6620	285.6290	284.3732	295.5165	283.8322	297.1240	284.5998
P_{10} (MW)	131.1614	204.8213	138.3372	134.3401	131.9480	193.0429	130.0000
P_{11} (MW)	98.4919	156.3835	176.2985	169.1616	241.4022	94.8783	168.7998
P_{12} (MW)	168.5389	172.5566	162.0961	100.6024	154.4301	172.7777	168.7998
P_{13} (MW)	214.1733	304.4306	215.7737	215.0265	214.3881	128.7132	214.7599
P_{14} (MW)	397.7859	304.6572	298.7209	390.9713	395.1666	395.9701	394.2794
P_{15} (MW)	393.1412	299.3851	394.1351	302.5692	390.4244	305.1189	394.2793
P_{16} (MW)	387.4290	394.0908	394.4072	392.6898	306.8219	393.6977	304.5197
P_{17} (MW)	485.7239	490.0678	490.9882	489.0061	496.8710	489.1214	489.2795
P_{18} (MW)	492.6118	493.0292	487.7777	486.1721	489.1898	489.4086	489.2795
P_{19} (MW)	512.9207	512.3289	507.5660	511.9301	516.8716	511.3859	511.2795
P_{20} (MW)	511.6986	511.2362	511.1204	516.4457	511.3494	511.8786	511.2794

(continued)

Table 6 (continued)

Generating unit	BBO-EM[a] PMB	MpBBO Lin.	Exp.	Inv.	Inv. Lin.	Log.	MpBBO-SQP Inv. Lin.
P_{21} (MW)	521.4651	523.2986	524.7963	535.3893	520.2215	522.9748	523.2794
P_{22} (MW)	524.7502	533.0381	533.9713	533.6980	519.2020	527.9580	523.2794
P_{23} (MW)	534.8721	525.2546	521.4333	519.7955	519.9372	516.3175	523.2794
P_{24} (MW)	522.3196	538.6236	521.6195	521.9118	525.9079	527.6274	523.2801
P_{25} (MW)	521.0463	524.3549	534.4174	528.5863	523.6342	528.7126	523.2800
P_{26} (MW)	522.2667	525.3645	529.7122	533.1516	529.7113	528.4108	523.2794
P_{27} (MW)	12.9535	11.5957	10.7071	18.0412	12.1148	16.0110	10.0000
P_{28} (MW)	16.2587	10.9665	15.4744	10.9164	11.1498	16.1873	10.0000
P_{29} (MW)	20.5544	10.1174	10.9433	10.0349	17.1487	14.6821	10.0000
P_{30} (MW)	96.5324	89.8871	90.8936	88.2024	91.6684	84.3462	89.7581
P_{31} (MW)	189.2199	181.5048	185.5420	185.7671	188.5363	189.4289	190.0000
P_{32} (MW)	188.3592	164.5573	189.2287	186.9412	189.1755	184.6063	190.0000
P_{33} (MW)	183.6452	169.0938	178.1893	188.7571	179.1483	188.4070	190.0000
P_{34} (MW)	171.3132	165.2316	164.1037	166.3106	163.2768	177.2380	164.8006
P_{35} (MW)	197.7943	182.2683	165.9807	170.2927	169.1983	171.9220	164.8002
P_{36} (MW)	193.9636	164.3957	166.6935	173.5405	170.3869	176.3794	164.8030
P_{37} (MW)	107.8912	103.0432	109.0115	109.1962	89.6925	91.7985	110.0000
P_{38} (MW)	99.5476	89.1103	98.4373	91.4707	99.2854	106.0449	110.0000
P_{39} (MW)	84.6454	99.4453	88.9657	88.6757	87.8194	105.4537	110.0000
P_{40} (MW)	508.2361	526.4697	509.9135	513.4386	520.6376	544.2860	511.2794

(continued)

Table 6 (continued)

Generating unit	BBO-EM[a]	MpBBO						MpBBO-SQP
	PMB	Lin.	Exp.	Inv.	Inv. Lin.	Log.		Inv. Lin.
$\sum P_i$ (MW)	10500.00	10500.00	10500.00	10500.00	10500.00	10500.00		10500.00
Best cost ($/h)	122762.2	122801.5	122577.9	122509.0	122642.4	122780.1		**121415.3**
Mean ($/h)	123337.0	123324.5	123080.7	123100.6	123057.2	123269.2		**122025.2**
Median ($/h)	123326.6	123324.9	123043.9	123023.2	123050.3	123251.5		**121989.1**
Std. Dev. ($/h)	268.1310	262.9107	243.7758	309.6664	**210.4609**	254.9956		343.0694
Avg. CPU time (s)	**12.53991**	13.68938	17.23036	16.83868	16.71288	13.52386		31.07582

[a]See the footnote of Table 2

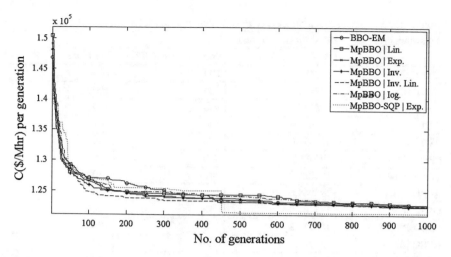

Fig. 8 The fitness curves of BBO versions for test case III

Table 6 shows the results obtained by BBO-EM, MpBBO and MpBBO-SQP after 50 trials. In this case, the inverse linear cooling strategy performed better than the others in terms of the average quality. Thus, MpBBO-SQP is executed with that cooling rate. It can be observed from Table 6 that the processing time of MpBBO increases as the solution quality increases. This phenomenon is clear with the exponential, inverse and inverse linear cooling rates. In this comparison, it can be clearly seen that MpBBO-SQP is the winner with no competitive. However, it consumes around 80–140 % additional CPU time. In this system, SQP phase was not activated unless reaching 90 % of G; as can be seen in Fig. 8. This approach will give enough chance to MpBBO phase to explore the search space, and it will save a significant amount CPU time. Table 7 shows an extended comparison between MpBBO-SQP and 17 different optimization algorithms presented in the literature. Again, MpBBO-SQP proved itself as a superior algorithm, which can detect better solutions with very small population size k and relatively few number of generations G.

5   Further Discussion

Based on the results presented, it can be concluded that the BBO solution quality can be steeply enhanced by reinforcing the migration stage through the Metropolis criterion of SA and fine tuning the elites through SQP. With this double hybridization approach, both the exploration and exploitation levels can be prettily balanced.

Table 7 Comparison between MpBBO-SQP and other algorithms for test case III

Algorithm type	Performance criteria		
	Best cost ($/h)	Population size (k)	Iterations no. (G)
CEP [16]	123488.3	100	1000
FEP [16]	122679.7	100	1000
MFEP [16]	122647.6	100	1000
IFEP [16]	122624.4	100	1000
PSO [64]	122324.0	40	500
PPSO [64]	121788.2	40	500
MPPSO [64]	122225.7	40	500
APPSO [64]	122044.6	40	500
DPSO [64]	122160.0	40	500
PSO-LRS [67]	122035.8	**20**	1000
NPSO [67]	121704.7	**20**	1000
NPSO-LRS [67]	121664.4	**20**	1000
CSO [68]	121461.7	30	1000
CDEMD [69]	121423.4	25	1000
ABC [12]	121441.0	800	200
FAPSO-NM [65]	121418.3	60	1000
BBO [13]	121510.8	500	**100**
MpBBO-SQP	**121415.3**	**20**	500

It has been found that the other properties of SA have also been transferred to MpBBO-SQP. Thus, the initial temperature $\check{T}_o$ and the cooling strategy type are very important factors that may affect on the overall performance. However, because BBO do not need much settings in the initialization stage, so MpBBO-SQP has less dependability on those SA parameters.

Also, it has been found that, although the internal searching loops l of SA is deactivated in MpBBO-SQP, it consumes huge CPU time. This is because of the second phase hybridization. It can be effectively decreased by reducing the number of elites that need to be fine-tuned through SQP and/or setting lower accuracy for the termination criterion (*in this study, the tolerance of* 10^{-7} *is used for SQP*). For example, the processing time of the 40-units test case can be steeply decreased from 31.08 s down to 8.13 s, with an optimal cost of 121478.0 $/h, if only the fitness is recycled through SQP instead of all the elites. This means that MpBBO-SQP can get very good solutions with the lowest CPU time. Additionally, it is good to initiate SQP after completing a big portion of the total iterations (*like 90% of G → as with the 40 units test case*), so that the algorithm can have more chance to explore the search space and at the same time the CPU usage can be effectively reduced.

6 Conclusions and Suggestions

A novel hybrid optimization algorithm called MpBBO-SQP is proposed in order to solve the economic dispatch problems of three different test cases. The objective of this hybridization is to have a good balance between the exploration and exploitation levels by combining the features of BBO, SA and SQP algorithms into one composite algorithm. The results obtained show the superiority of MpBBO-SQP as compared with many different algorithms presented in the literature. It can be said that the proposed MpBBO-SQP algorithm has high immunity against trapping into local optimums, but the main drawback of this hybrid algorithm is its gluttony in the CPU usage. However, the root of this issue is already addressed and thus it can be easily solved by reducing the number of individuals that need to be fine-tuned through SQP, and also by reducing the tolerance setting of SQP termination criterion.

Many options can be suggested to increase the performance of MpBBO-SQP in terms of solution quality and processing speed. One of these options is finding the optimal settings by conducting some sensitivity analyses. Moreover, it is good to test MpBBO-SQP with adaptive cooling strategies. Furthermore, the mutation and migration rates of BBO can be replaced with others presented in the literature. Also, oppositional and blended BBO models (OBBO and BBBO) can be considered as new phases for more advanced hybrid algorithms.

References

1. Grainger, J.J., Stevenson, W.D.: Power System Analysis. McGraw-Hill Education, New York (1994)
2. Lee, F., Breipohl, A.: Reserve constrained economic dispatch with prohibited operating zones. IEEE Trans. Power Syst. **8**(1), 246–254 (1993)
3. Park, J., Kim, Y., Eom, I., Lee, K.: Economic load dispatch for piecewise quadratic cost function using hopfield neural network. IEEE Trans. Power Syst. **8**(3), 1030–1038 (1993)
4. Bakirtzis, A., Petridis, V., Kazarlis, S.: Genetic algorithm solution to the economic dispatch problem. IEE Proc. Gener. Transm. Distrib. **141**(4), 377–382 (1994)
5. Farag, A., Al-Baiyat, S., Cheng, T.C.: Economic load dispatch multiobjective optimization procedures using linear programming techniques. IEEE Trans. Power Syst. **10**(2), 731–738 (1995)
6. Wood, A.J., Wollenberg, B.F.: Power Generation, Operation, and Control, 2nd edn. Wiley-Interscience, New York (1996)
7. Saadat, H.: Power System Analysis. McGraw-Hill Series in Electrical and Computer Engineering. WCB/McGraw-Hill, Singapore (1999)
8. Abido, M.: Environmental/economic power dispatch using multiobjective evolutionary algorithms. IEEE Trans. Power Syst. **18**(4), 1529–1537 (2003)
9. Gaing, Z.-L.: Particle swarm optimization to solving the economic dispatch considering the generator constraints. IEEE Trans. Power Syst. **18**(3), 1187–1195 (2003)
10. Husain, A.: Electrical Power Systems, 5th edn. CBS Publishers & Distributors, New Delhi, India (2007)
11. Mahor, A., Prasad, V., Rangnekar, S.: Economic dispatch using particle swarm optimization: a review. Renew. Sust. Energy Rev. **13**(8), 2134–2141 (2009)

12. Hemamalini, S., Simon, S.P.: Artificial bee colony algorithm for economic load dispatch problem with non-smooth cost functions. Electr. Power Compon. Syst. **38**(7), 786–803 (2010)
13. Mohammed, Z., Talaq, J..: Economic dispatch by biogeography based optimization method. In: 2011 International Conference on Signal, Image Processing and Applications, vol. 21, pp. 161–165. IACSIT Press, Singapore (2011)
14. Karthikeyan, V., Senthilkumar, S., Vijayalakshmi, V.: A new approach to the solution of economic dispatch using particle swarm optimization with simulated annealing. Int. J. Comput. Sci. Appl. (IJCSA) **3**(3), 37–49 (2013)
15. Fister Jr., I., Yang, X.-S., Fister, I., Brest, J., Fister, D.: A brief review of nature-inspired algorithms for optimization. Elektrotehniški Vestnik **80**(3), 1–7 (2013)
16. Sinha, N., Chakrabarti, R., Chattopadhyay, P.: Evolutionary programming techniques for economic load dispatch. IEEE Trans. Evol. Comput. **7**(1), 83–94 (2003)
17. Rahmat, N., Musirin, I.: Differential evolution ant colony optimization (DEACO) technique in solving economic load dispatch problem. In: IEEE International Conference on Power Engineering and Optimization Conference (PEDCO). Melaka, Malaysia, Jun. 2012, pp. 263–268 (2012)
18. Simon, D.: Biogeography-based optimization. IEEE Trans. Evol. Comput. **12**(6), 702–713 (2008)
19. Pattnaik, S., Lohokare, M., Devi, S.: Enhanced biogeography-based optimization using modified clear duplicate operator. In: Second World Congress on Nature and Biologically Inspired Computing (NaBIC), Dec. 2010, pp. 715–720 (2010)
20. Gong, W., Cai, Z., Ling, C.X., Li, H.: A real-coded biogeography-based optimization with mutation. Appl. Math. Comput. **216**(9), 2749–2758 (2010)
21. Gong, W., Cai, Z., Ling, C.: DE/BBO: a hybrid differential evolution with biogeography-based optimization for global numerical optimization. Soft Comput. **15**(4), 645–665 (2010)
22. Lohokare, M., Pattnaik, S., Panigrahi, B., Das, S.: Accelerated biogeography-based optimization with neighborhood search for optimization. Appl. Soft Comput. **13**(5), 2318–2342 (2013)
23. Alroomi, A.R., Albasri, F.A., Talaq, J.H.: Solving the associated weakness of biogeography-based optimization algorithm. Int. J. Soft Comput. **4**(4), 1–20 (2013). http://dx.doi.org/10.5121/ijsc.2013.4401
24. Hetzer, J., Yu, D., Bhattarai, K.: An economic dispatch model incorporating wind power. IEEE Trans. Energy Convers. **23**(2), 603–611 (2008)
25. Chaturvedi, K., Pandit, M., Srivastava, L.: Self-organizing hierarchical particle swarm optimization for nonconvex economic dispatch. IEEE Trans. Power Syst. **23**(3), 1079–1087 (2008)
26. Pavri, R., Moore, G.D.: Gas turbine emissions and control. GE Energy Services, Atlanta, GA, Technical Report GER-4211, Mar. 2001. http://site.ge-energy.com/prod_serv/products/tech_docs/en/downloads/ger4211.pdf
27. Simon, D.: Evolutionary Optimization Algorithms: Biologically-Inspired and Population-Based Approaches to Computer Intelligence. Wiley, Hoboken, New Jersey (2013)
28. Eiben, A., Smith, J.: Introduction to Evolutionary Computing. Natural Computing Series. Springer, Berlin (2003)
29. Rao, S.S.: Engineering Optimization: Theory and Practice, 4th edn. Wiley, Hoboken, New Jersey (2009)
30. Nocedal, J., Wright, S.: Operations Research and Financial Engineering. Numerical Optimization, 2nd edn. Springer, New York (2006)
31. Kirkpatrick, S., Gelatt, C.D., Vecchi, M.P.: Optimization by simulated annealing. Science **220**(4598), 671–680 (1983)
32. Černý, V.: Thermodynamical approach to the traveling salesman problem: an efficient simulation algorithm. J. Optim. Theory Appl. **45**(1), 41–51 (1985)
33. Metropolis, N., Rosenbluth, A.W., Rosenbluth, M.N., Teller, A.H., Teller, E.: Equation of state calculations by fast computing machines. J. Chem. Phys. **21**(6), 1087–1092 (1953)
34. Lomolino, M.V., Riddle, B.R., Brown, J.H.: Biogeography, 3rd edn. Sinauer Associates, Sunderland, Mass (2009)

35. Jones, R.: Biogeography: Structure, Process, Pattern, and Change Within the Biosphere. Hulton Educational Publications Limited (1980)
36. Cox, C.B., Moore, P.D.: Biogeography: An Ecological and Evolutionary Approach. Blackwell Scientific Publications, Oxford England Cambridge, Mass (1993)
37. MacArthur, R.H., Wilson, E.O.: An equilibrium theory of insular zoogeography. Evolution **17**(4), 373–387 (1963)
38. MacArthur, R.H., Wilson, E.O.: The Theory of Island Biogeography. Landmarks in Biology Series. Princeton University Press, Princeton (1967)
39. Nierenberg, W.A. (ed.): Encyclopedia of Environmental Biology, Three-Volume Set: 1–3, 1st edn. Academic Press (1995)
40. Myers, A.A., Giller, P.S. (eds.): Analytical Biogeography: An Integrated Approach to the Study of Animal and Plant Distributions, 1st edn. Chapman and Hall, London (1990)
41. Cody, M.L.: Plants on Islands: Diversity and Dynamics on a Continental Archipelago. University of California Press, Berkeley, Calif (2006)
42. MacArthur, R.H., Connell, J.H.: The Biology of Populations. Wiley, New York (1966)
43. Losos, J.B., Ricklefs, R.E.: The Theory of Island Biogeography Revisited. Princeton University Press, Princeton (2010)
44. MacArthur, R.H.: Geographical Ecology: Patterns in the Distribution of Species. Harper & Row Publishers Inc., New York (1972)
45. Igelnik, B., Simon, D.: The eigenvalues of a tridiagonal matrix in biogeography. Appl. Math. Comput. **218**(1), 195–201 (2011)
46. MacDonald, G.M.: Biogeography: Introduction to Space, Time, and Life. Wiley, New York (2003)
47. Simon, D.: A probabilistic analysis of a simplified biogeography-based optimization algorithm. Evol. Comput. **19**(2), 167–188 (2011)
48. Alroomi, A.R., Albasri, F.A., Talaq, J.H.: A comprehensive comparison of the original forms of biogeography-based optimization algorithms. Int. J. Soft Comput. Artif. Intell. Appl. **2**(5/6), 11–30 (2013). http://dx.doi.org/10.5121/ijscai.2013.2602
49. Ma, H., Simon, D.: Blended biogeography-based optimization for constrained optimization. Eng. Appl. Artif. Intel. **24**(March), 517–525 (2011)
50. Simon, D.: The Matlab Code of Biogeography-Based Optimization, Aug. 2008. http://academic.csuohio.edu/simond/bbo/. Accessed 1 Feb 2013
51. Simon, D., Ergezer, M., Du, D.: Population distributions in biogeography-based optimization algorithms with elitism. In: IEEE International Conference on in Systems, Man and Cybernetics, SMC 2009, Oct. 2009, pp. 991–996 (2009)
52. Nourani, Y., Andresen, B.: A comparison of simulated annealing cooling strategies. J. Phys. A: Math. Gener. **31**(41), 8373–8385 (1998)
53. Lundy, M., Mees, A.: Convergence of an annealing algorithm. Math. Program. **34**(1), 111–124 (1986)
54. Szu, H., Hartley, R.: Fast simulated annealing. Phys. Lett. A **122**, 157–162 (1987)
55. Geman, S., Geman, D.: Stochastic relaxation, Gibbs distributions, and the Bayesian restoration of images. IEEE Trans. Pattern Anal. Mach. Intell. **PAMI-6**(6), 721–741 (1984)
56. Victoire, T.A., Jeyakumar, A.: Hybrid PSO-SQP for economic dispatch with valve-point effect. Electr. Power Syst. Res. **71**(1), 51–59 (2004)
57. Albasri, F., Alroomi, A., Talaq, J.: Optimal coordination of directional overcurrent relays using biogeography-based optimization algorithms. IEEE Trans. Power Delivery **30**(4), 1810–1820 (2015)
58. Alsumait, J., Sykulski, J., Al-Othman, A.: A hybrid GA-PS-SQP method to solve power system valve-point economic dispatch problems. Appl. Energy **87**(5), 1773–1781 (2010)
59. Noghabi, A., Sadeh, J., Mashhadi, H.: Considering different network topologies in optimal overcurrent relay coordination using a hybrid GA. IEEE Trans. Power Delivery **24**(4), 1857–1863 (2009)
60. Bedekar, P., Bhide, S.: Optimum coordination of directional overcurrent relays using the hybrid GA-NLP approach. IEEE Trans. Power Delivery **26**(1), 109–119 (2011)

61. Yang, H.-T., Yang, P.-C., Huang, C.-L.: Evolutionary programming based economic dispatch for units with non-smooth fuel cost functions. IEEE Trans. Power Syst. **11**(1), 112–118 (1996)
62. Park, J.-B., Lee, K.-S., Shin, J.-R., Lee, K.: A particle swarm optimization for economic dispatch with nonsmooth cost functions. IEEE Trans. Power Syst. **20**(1), 34–42 (2005)
63. Chiang, C.-L.: Improved genetic algorithm for power economic dispatch of units with valve-point effects and multiple fuels. IEEE Trans. Power Syst. **20**(4), 1690–1699 (2005)
64. Chen, C., Yeh, S.: Particle swarm optimization for economic power dispatch with valve-point effects. In: Transmission Distribution Conference and Exposition: Latin America, TDC '06. IEEE/PES, Aug. 2006, pp. 1–5 (2006)
65. Niknam, T.: A new fuzzy adaptive hybrid particle swarm optimization algorithm for non-linear, non-smooth and non-convex economic dispatch problem. Appl. Energy **87**(1), 327–339 (2010)
66. Niknam, T., Mojarrad, H.D., Meymand, H.Z.: A novel hybrid particle swarm optimization for economic dispatch with valve-point loading effects. Energy Convers. Manag. **52**(4), 1800–1809 (2011)
67. Selvakumar, A., Thanushkodi, K.: A new particle swarm optimization solution to nonconvex economic dispatch problems. IEEE Trans. Power Syst. **22**(1), 42–51 (2007)
68. Selvakumar, A.I., Thanushkodi, K.: Optimization using civilized swarm: solution to economic dispatch with multiple minima. Electr. Power Syst. Res. **79**(1), 8–16 (2009)
69. dos Santos Coelho, L., Souza, R.C.T., Mariani, V.C.: Improved differential evolution approach based on cultural algorithm and diversity measure applied to solve economic load dispatch problems. Math. Comput. Simul. **79**(10), 3136–3147 (2009)

Gravitational Search Algorithm Applied to the Cell Formation Problem

Manal Zettam and Bouazza Elbenani

Abstract Group technology is a concept that emerged in the manufacturing field almost seventy years ago. Since then, group technology has been widely applied by means of a cellular manufacturing philosophy application called the cell formation problem. In this paper, we focus on adapting the discrete gravitational search algorithm to the cell formation problem. The mathematical model and the discrete gravitational search algorithm stages are detailed thereafter. To evaluate the algorithm's performance, thirty-five tests were carried out on widely used benchmarks. The results obtained were satisfactory to confirm successful adaptation of the gravitational search algorithm. Indeed, the algorithm reached thirty best values of benchmarks obtained by previous algorithms. The algorithm also outperformed the best-known solution of one benchmark.

Keywords Cell formation problem · Independent movement search phase · Gravitational search algorithm · Local search · Nature inspired metaheuristics

1 Introduction

In cellular manufacturing philosophy, a system consists of interacting sets of machines and parts. The cell formation problem (CFP) consists of subdividing the machines and parts sets into autonomous cells in such a way that minimizes inter-cell movements while maximizing intra-cell movements.

A survey of methods used to solve the CFP can be found in [22]. This survey briefly introduces the method categories used to solve the CFP, such as Cluster Analysis, Graph Partitioning Approaches, Mathematical Programming Methods, Heuristic and Metaheuristic Algorithms and Artificial Intelligence Methodologies.

M. Zettam (✉) · B. Elbenani
Department of Computer Science, Mohammed V University, Rabat, Morocco
e-mail: zettammanal@gmail.com

B. Elbenani
e-mail: elbenani@fsr.ac.ma

© Springer International Publishing Switzerland 2016
X.-S. Yang (ed.), *Nature-Inspired Computation in Engineering*,
Studies in Computational Intelligence 637, DOI 10.1007/978-3-319-30235-5_12

Cluster Analysis methods are recognized as cell formation tools. Cluster Analysis methods aim to group entities into clusters such that entities have high-level interactions within a cluster and low-level interactions between clusters. Cluster Analysis methods are categorized into the following three method categories:

- Array-based clustering,
- Hierarchical clustering,
- Non-hierarchical clustering techniques.

Further details about those three method categories can be found in [22].

For Graph Partitioning Approaches, a network or graph is employed to represent the cell formation problem. Vertices of the representative graph or network could be a machine or a part, while the parts processing are represented by edges. Cluster Analysis methods and Graph Partitioning Approaches are known as one objective methods, meaning that only minimization of inter-cell movements is taken into account. To tackle the weakness of the previous cited methods, Mathematical Programming Methods were used to formulate the cell formation problem.

Mathematical programming formulations for CFP are either nonlinear or linear integer programming problems. The mathematical programming formulations enable incorporation of setup and processing time, the use of multiple identical machines, and ordered sequences as well as other constraints. Nonetheless, incorporating constraints into the model increases the intractability of the cell formation problem i.e. CFP is a NP-complete(hard) problem [5]. Therefore, a great number of Heuristics, Metaheuristics and Artificial Intelligence Methodologies have been employed to solve the CFP.

Artificial Intelligence Methodologies are classified under two categories: neural networks and fuzzy methodologies. Neural networks (NNs) are well-known for their robustness and high capacity for adaption; therefore, NNs were extensively applied to CFP. Several types of NNs were applied to the CFP, such as Hopfield network [35], self-organizing map [32], adaptive resonance theory [4] and transiently chaotic neural network [28].

Heuristics and Metaheuristics provide good solutions in reasonable computational time for large-size NP-hard problems. Unlike metaheuristics, which are applicable to large-scale problems, heuristics are designed for a specific problem. Mukattash et al. [21] proposed three specific heuristics to solve the CFP. Tabu search (TS) [17], simulated annealing (SA) [29], genetic algorithms (GAs) [13], particle swarm optimization (PSO) [25] and ant colony optimization (ACO) [18] are well-known metaheuristics applied to the CFP.

In this chapter, we detailed a number of changes made to the discrete gravitational search algorithm (DGSA) performed by [7]. These changes were performed to enhance the previously introduced DGSA. The original DGSA consists mainly of two phases. The first phase, called the Dependent Movement Search Operator Phase (DMSOP), uses a Path-Relinking algorithm to find a new solution. The second DSGA phase, called the Independent Movement Operator Phase (IMSOP), is a modified local search. Since the trade-off between exploration and exploitation was only performed by means of a set consisting of the best solutions of the population, exploration takes the lead. Indeed, the Path-Relinking algorithm and the local search use only mutation operators. The algorithm will not suit the structure of most combinatorial problems. Consequently, a crossover is used with the main goal of balancing the exploration and exploitation strategies. In that way, the proposed DGSA is more likely to reach better solutions in less computational time.

The remainder of this paper is organized as follows: in the second section, both the mathematical model and the CFP are introduced, the third section presents a brief survey of nature-inspired metaheuristics, the fourth section details the original GSA, the fifth section details the changes applied to DGSA for successful adaptation of the algorithm to CFP, the sixth section provides computational results and a comparative study, and the conclusion contains final remarks and suggestions for future works.

2 The Cell Formation Problem

2.1 The Incidence Matrix

The CFP is often attributed to Burbidge's work detailed in [2]. Burbidge proved that the CFP could be reduced to a functional grouping of parts and machines based on a binary matrix called the incidence matrix. The incidence matrix is a 0–1 machine-part incidence matrix. As pointed out, the main aim of the CFP is to reduce the inter-cell movements and increase the intra-cell movements, which could be done by arranging the rows and the columns of the incidence matrix in such a manner as to form a diagonal bloc matrix. The method performances used to solve the CFP were evaluated by one of the performance measures detailed in [11]. In this work the Grouping Efficacy measure was utilized. The Grouping Efficacy measure is defined as follows:

$$Eff = \frac{a - a_1^{Out}}{a + a_0^{In}} = \frac{a_1^{In}}{a + a_0^{In}} \tag{1}$$

where a is the total number of 1 in the incidence matrix, a_0^{In} is the number of zeros within a cell which are named void elements, a_1^{Out} is the number of 1 out of all cells called exceptional elements, a_0^{In} and is the number 1 of within a given cell.

Figure 1 shows an example of an incidence matrix and the solution matrix obtained by applying a solving method. In Fig. 1b the Grouping Efficacy is equal to:

$$Eff = \frac{20 - 1}{20 + 8} = 67.86\%$$

(a)

Parts	1	2	3	4	5	6	7	8	9	10	11
1	1	1	0	0	0	1	0	0	0	0	0
2	0	1	0	0	0	1	0	0	1	0	0
3	1	0	0	0	0	0	1	0	0	0	1
4	0	0	1	0	0	0	1	0	0	0	0
5	0	0	1	1	0	0	0	0	0	0	1
6	0	0	0	0	1	0	0	0	0	1	0
7	0	0	0	0	1	0	0	1	0	1	0

(Machines label on left side, rows 1–7)

(b)

Parts	4	5	8	10	1	2	6	9	3	7	11
6	0	1	0	1	0	0	0	0	0	0	0
7	0	1	1	1	0	0	0	0	0	0	0
1	0	0	0	0	1	1	1	0	0	0	0
2	0	0	0	0	0	1	1	1	0	0	0
3	0	0	0	0	1	0	0	0	0	1	1
4	0	0	0	0	0	0	0	0	1	0	0
5	0	0	0	0	0	0	0	0	1	1	1

(Machines label on left side)

Fig. 1 An incidence matrix example

2.2 The Mathematical Model

The mathematical programming model formulated below is similar to the one presented in [8]:

$$Max\ Eff = \frac{\sum_{k=1}^{C} \sum_{i=1}^{M} \sum_{j=1}^{P} a_{ij} x_{ik} y_{ik}}{a + \sum_{k=1}^{C} \sum_{i=1}^{M} \sum_{j=1}^{P} (1 - a_{ij}) x_{ij} y_{ik}} \quad (2)$$

Subject to:

$$\sum_{k=1}^{C} x_{ik} = 1, \quad i = 1, ..., M \quad (3)$$

$$\sum_{k=1}^{C} y_{jk} = 1, \quad j = 1, ..., P \quad (4)$$

$$\sum_{i=1}^{M} x_{ik} \geq 1, \quad k = 1, ..., C \quad (5)$$

$$\sum_{j=1}^{P} y_{jk} \geq 1, \quad k = 1, ..., C \quad (6)$$

$$x_{ik} = 0\ \text{or}\ 1, \quad i = 1, ..., M, \quad k = 1, ..., C \quad (7)$$

$$y_{jk} = 0\ \text{or}\ 1, \quad j = 1, ..., P, \quad k = 1, ..., C \quad (8)$$

where M, P, C denotes respectively the number of machines, the number of parts and the number of cells. x_{ik} and y_{ik} are two binary variables defined as below:
For each $i = 1, ..., M, k = 1, ..., C$

$$x_{ik} = \begin{cases} 1, & \text{if the ith machine belongs to the kth cell} \\ o & \text{otherwise} \end{cases}$$

For each $j = 1, ..., P, k = 1, ..., C$

$$y_{jk} = \begin{cases} 1, & \text{if the jth machine belongs to the kth cell} \\ o & \text{otherwise} \end{cases}$$

And where

$$a_1^{Out} = a - \sum_{k=1}^{C} \sum_{i=1}^{M} \sum_{j=1}^{P} a_{ij} x_{ik} y_{jk} \qquad (9)$$

$$a_0^{In} = \sum_{k=1}^{C} \sum_{i=1}^{M} \sum_{j=1}^{P} (1 - a_{ij}) x_{ik} y_{jk} \qquad (10)$$

The (3) and (4) equations ensure that each part and each machine must be assigned to only one cell. The (5) and (6) equations guarantee that each cell contains at least one part and one machine. The (7) and (8) equations denote that decision variables are binary. In the computational experiments, the number of cells, denoted k, was fixed for each problem to its value in the best-known solution reported in the literature.

3 Nature-Inspired Metaheuristics

Nature-inspired metaheuristics have gained great notoriety in recent decades. The first nature-inspired metaheuristic was defined by Holland [14, 15]. The success of solving difficult problems encountered by GA, has promoted the application of nature-inspired metaheuristics. Some of well-known nature-inspired metaheuristics are listed below:

- Artificial Immune Algorithm (AIA) [10]. AIA is inspired by the human immune system,
- Ant Colony Optimization (ACO) [26]. ACO is inspired by the foraging behavior of ants,
- Particle Swarm Optimization (PSO) [3]. PSO is inspired by the foraging behavior of swarms of birds,
- Harmony Search (HS) [33]. HS is inspired from a musician's improvisation,
- Bacteria Foraging Optimization (BFO) [23]. BFO is inspired from the behavior of bacteria,
- Shuffled Frog Leaping (SFL) [9]. SFL is inspired from communication amongst frogs,
- Artificial Bee Colony (ABC) [1]. ABC is inspired from the foraging behavior of a honeybee,
- Biogeography-Based Optimization (BBO) [27]. BBO is inspired from of immigration and emigration processes of various species,
- Cuckoo Search (CS) [34]. CS is inspired from the obligate brood parasitism of cuckoo species.
- Gravitational Search Algorithm (GSA) [24]. CSA is inspired by Newton's law of gravitation.

All of these nature-inspired population-based optimization methods have been successfully applied to optimization problems, particularly those regarding scheduling
and assignment. We focus our attention on the Gravitational search algorithm, which
also has been employed to solve optimization problems, especially scheduling and
assignment problems [6, 30]. The effectiveness of GSA and its variants for scheduling and assignment problems was an impetus to resolve the CFP by using GSA.

4   The Original Gravitational Search Algorithm

The Gravitational Search Algorithm (GSA) was first proposed by [24]. The GSA is
considered a nature-inspired algorithm based on the Newton's Law of Gravity and the
Law of Motion. In the GSA, each agent is considered an object and its performance
is measured by its mass. The gravity force generated by the attraction between the
objects, causes a global movement towards the objects with a heavier mass. After a
while, the masses will be attracted by the heaviest mass that represents the optimum
solution in the search space.

Now, a system of N agents (masses) is considered. The ith agent position is defined
as follows:

$$X_i = \left(x_i^1, ..., x_i^d, ..., x_i^n\right) \quad for \; i = 1, 2, ..., N \tag{11}$$

where x_i^d represents the ith agent position in the dth dimension.

At a specific time t, the force acting on mass i from mass j is defined by the
following equation:

$$F_{ij}^d(t) = G(t) \frac{M_{pi}(t) \times M_{aj}(t)}{R_{ij}(t) + \xi} \left(x_j^d(t) - x_i^d(t)\right) \tag{12}$$

where $M_{aj}(t)$ represents the active gravitational mass related to agent j, $M_{pi}(t)$ represents the passive gravitational mass related to agent i, $G(t)$ represents the gravitational
constant at time t, ξ represents a small constant, and $R_{ij}(t)$ represents the Euclidian
distance between two agents i and j:

$$R_{ij}(t) = \left\| X_i(t), X_j(t) \right\|_2 \tag{13}$$

To give a stochastic characteristic to the GSA, the total force that acts on agent i
in a dimension d is supposed to be a randomly weighted sum of dth components of
the forces exerted from other agents:

$$F_i^d(t) = \sum_{j=1, j \neq i}^{N} rand_j F_{ij}^d(t) \tag{14}$$

To improve the performance of the GSA, the authors of [?] define *kbest* as a function of time, with the initial value k_0 at the beginning and decreasing linearly until there will be just one agent within the *Kbest* set (the set of better solutions). Consequently, the Eq. (24) was modified as follows:

$$\bar{F}_i^d(t) = \sum_{j \in Kbest, j \neq i}^{N} rand_j F_{ij}^d(t) \tag{15}$$

where $rand_j$ denotes a random number in the interval $[0, 1]$. Thus, by the Law of Motion, the acceleration of the agent i at a specific time t, and in direction dth, $a_i^d(t)$ is defined by:

$$a_i^d(t) = \frac{\bar{F}_i^d(t)}{M_{ii}(t)} \tag{16}$$

where $M_{ii}(t)$ denotes the ith agent inertial mass.

Moreover, the next velocity of an agent is represented by a fraction of its current velocity added to its acceleration. Therefore, its position and velocity could be calculated as follows:

$$v_i^d(t+1) = rand_i \times v_i^d(t) + a_i^d(t) \tag{17}$$

$$x_i^d(t+1) = x_i^d(t) + v_i^d(t+1) \tag{18}$$

where $rand_i$ denotes a uniform random variable in the interval $[0, 1]$. This random number is used to give a randomized characteristic to the GSA.

The gravitational constant G is initialized at the beginning of the algorithm and its value reduced during the search process to control the search accuracy. In other words, G is a function of the initial value (G_0) and time (t):

$$G(t) = G(G_0, t) \tag{19}$$

The gravitational and the inertia masses are calculated by the fitness evaluation. A heavier mass means a more efficient agent. Therefore, better agents move more slowly and have a higher attraction. Assuming the equality of the gravitational and inertia mass, the mass's values are calculated using the fitness map. The gravitational and inertial masses are updated by the following equations:

$$M_{ai} = M_{pi} = M_{ii} = M_i \quad i = 1, 2, ..., N \tag{20}$$

$$m_i(t) = \frac{fit_i(t) - worst(t)}{best(t) - worst(t)} \tag{21}$$

$$M_i(t) = \frac{m_i(t)}{\sum_{j=1}^{N} m_j(t)} \tag{22}$$

where $fit_i(t)$ represents the agent i fitness value at time t, $worst(t)$ and $best(t)$ are defined for a minimization problem as follows:

$$best(t) = \max(fit_j(t))_{j\in\{1,...,N\}} \qquad (23)$$

$$worst(t) = \min(fit_j(t))_{j\in\{1,...,N\}} \qquad (24)$$

In this paper, the Grouping Efficiency is considered the fitness for the GSA. The fitness is defined as follows:

$$fit_i(t) = Eff = \frac{a - a_1^{Out}}{a + a_0^{In}} = \frac{a_1^{In}}{a + a_0^{In}} \qquad (25)$$

5 The Discrete Gravitational Search Algorithm

The DGSA was first introduced in [7]. As pointed out earlier in this paper, the DGSA has two search components named DMSOP and IMSOP. The Fig. 2 summarizes the original DGSA algorithm.

In the original DGSA, the DMSOP consists of a Path-Relinking algorithm. In our proposal, the DMSOP involves an acceleration vector and a crossover operator. The crossover operator returns a new position in the solution space based on both the agent's current position and the acceleration vector. The acceleration vector is calculated by means of the formula (16).

Fig. 2 The gravitational search algorithm

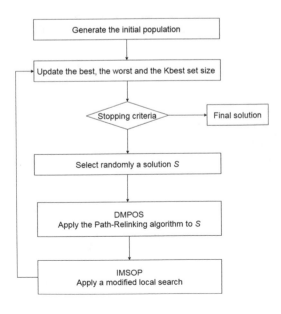

In the original DGSA, the IMSOP is a modified version of a local search algorithm that takes the velocity as a stopping criterion. The velocity is calculated by means of the Eq. (16). In our proposal, the DMSOP involves two mutation operators.

5.1   The Dependent Movement Search Operator Phase

The DMSOP involves both a set of good solutions and an algorithm called the Multi-Target Movement in Neighborhood Space (MTMNS). The set of good solutions is called the *Kbest* set. The *Kbest* set is also referred to as the target solutions set. The MTMNS algorithm was defined by [7] as an algorithm which applies l_i iterative changes to a starting solution S so that the distance between the starting solution S and a target solution S_i is minimized. The S_i represents the ith element of the target solutions set and l_i variable denotes the DML of S_i. The DML is the dependent movement length representing the greater element of the acceleration vector. The Algorithm (1) shows the template used for the DMSOP. The distance between two solutions S and S_i is denoted $Dist(S, S_i)$ This distance represents the number of different elements of two solutions.

Algorithm (1): Dependent Movement Search Operator Phase (DMSOP)Template
Inputs:
    the starting solution S;
    the k target solutions set $\{S_1, S_2, ...S_k\}$;
    the DML set $\{l_1, l_2, ...l_k\}$;
For $i = 1$ to k
$cpt = 1$;
While $Dist(S, S_i) \neq 0$ and $cpt \leq l_i$
    $S = S \otimes m$ /*apply an operator to decrease $Dist(S, S_i)$ */
    $cpt + +$;
    End While
End For
Output: The best solution found in trajectory

In our work, we have slightly changed the original DMSOP. Indeed, we have replaced the mutation operator with a crossover. Furthermore, the uniform crossover used was influenced by an accelerator vector instead of the DML. The Algorithm (2) shows the template used for our DMSOP. The changes brought do not only balance the exploration and exploitation strategies, but also reduce the computational time and complexity of the algorithm.

Algorithm (2): Dependent Movement Search Operator Phase Template
Inputs:
    the starting solution S;
    the K target solutions set $\{S_1, S_2, ..., S_k\}$;

the acceleration vector set $\{a_1, a_2, ..., a_k\}$;
For $i = 1 \, to \, k$
    Generate a vector v according to a_i
    $S = S \oplus S_i$/* Apply a crossover */
End For
Output: The best solution found in trajectory

5.2 The Independent Movement Search Operator Phase

As described above in the introduction, the IMSOP is a modified local search algorithm. In this work, we performed an iterative improvement algorithm that involves two movement operators. The operators are *add/drop* and *swap*. The IMSOP starts with a solution S and iteratively seeks for a better solution S' by applying one of the two previous cited operators with a probability equal to. Moreover, the neighborhood size is equal to $\lfloor \frac{M}{4} \rfloor$, where M represents the number of machines and $\lfloor x \rfloor$ denotes the integer part of x. The algorithm (3) shows the template we used for the IMSOP.

Algorithm (3): The Independent Movement Search Operator Phase Template
Input:
    The starting solution S;
    IML as the maximum movement length;
    The neighborhood size *Neighbors*;
    $S'' = S$;
    iter = 1;
Repeat
  For $i = 1$ to *Neighbors*
  Generate a random r number between 0 and 1;
  If $r \leq 0.5$ Then
      S'=Apply the *add/drop* operator to S;
Else
      S'=Apply the *swap* operator to S;
  End If
  If S' is better than S'' Then
  $S'' = S'$;
  End If
  End For
  $S = S''$;
  iter + +;
Until iteration==IML OR other stopping criteria are met;
Output: Final solution.

Table 1 DGSA parameters

Parameter	Value
The population size	$3M$
The size of *Kbest* set	$3M - 1$
The initial value of G	100

6 The Computational Results

In this section, we present test results carried out on 35 widely-used benchmarks. The proposed heuristic was coded in JAVA language and run on a personal computer with an Intel Core i5-2540M 2.60 GHz processor. For each benchmark, we ran the program 10 times. The obtained results were compared to the best-known solutions in the problem literature. Table 2 contains the best, worst and average grouping efficacies for the DGSA. Moreover, Table 2 reports the machine number, part number, cell number and the gap. Table 2 also contains the best-known grouping efficacies and their references.

The gap is a measure which represents the difference between the obtained results of an approach, denoted A, and our approach, denoted B. The gap is calculated as follows:

$$gap(\%) = \frac{the\ average\ efficacy\ A - the\ average\ efficacy\ B}{the\ average\ efficacy\ A} \times 100 \qquad (26)$$

One of the most relevant parameters in the DGSA is the size of the *Kbest* set. Indeed, to find a trade-off between exploration and exploitation, the *kbest* set size is set at a greater value equal to the initial population size minus one. This value is then decreased by means of a linear function represented by the 27th equation. In our work, we fixed the size of the population at thrice the number of machines. The initial value of the gravitational constant is denoted G_0. Table 1 contains the parameters and their values.

$$Size_{Kbest}(t + 1) = Size_{kbest}(t) - 1 \qquad (27)$$

The gap contained in Table 2 is obtained by comparing the best values of grouping efficacy, obtained by the proposed DGSA, and the best-known solution from the literature. Table 2 shows that the proposed DGSA outperforms the best-known solution for 1 problem (P33). Table 2 also shows that the proposed DGSA reaches the grouping efficacy values for 29 problems. The results were satisfactory and show that the DGSA was successfully adapted to the CFP.

Figure 3 indicates the gap value for the thirty-five instances. From Table 2 and Fig. 3, we concluded that, even though the results were satisfactory, the algorithm did not reach the best-known values of four benchmarks (P18, P21, P27 and P31). The algorithm is also not stable for five problems (P18, P26, P30, P32 and P33).

Table 2 Computational results

Problem	M	P	C	DGSA			Best known solution	Reference	Gap (%)
				Worst Eff	Best Eff	Average Eff			
P1	5	7	2	82.35	82.35	82.35	82.35	[31]	0
P2	5	7	2	69.57	69.57	69.57	69.57	[31]	0
P3	5	18	2	79.59	79.59	v79.59	79.59	[31]	0
P4	6	8	2	76.92	76.92	76.92	76.92	[31]	0
P5	7	11	5	60.87	60.87	v60.87	60.87	[31]	0
P6	7	11	4	70.83	70.83	70.83	70.83	[31]	0
P7	8	12	4	69.44	69.44	69.44	69.44	[31]	0
P8	8	20	3	85.25	85.25	85.25	85.25	[31]	0
P9	8	20	2	58.72	58.72	58.72	58.72	[31]	0
P10	10	10	5	75.00	75.00	75.00	75.00	[12]	0
P11	10	15	3	92.00	92.00	92.00	92.00	[12]	0
P12	14	24	7	72.06	72.06	72.06	72.06	[16]	0
P13	14	24	7	71.83	71.83	71.83	71.83	[16]	0
P14	16	24	8	53.26	53.26	53.26	53.26	[31]	0
P15	16	30	6	69.53	69.53	69.53	69.53	[8]	0
P16	16	43	8	57.53	57.53	57.53	57.53	[31]	0
P17	18	24	9	57.73	57.73	57.73	57.73	[31]	0
P18	20	20	5	43.07	43.17	43.12	43.45	[19]	0.64
P19	20	23	7	50.81	50.81	50.81	50.81	[16]	0
P20	20	35	5	77.91	77.91	77.91	77.91	[31]	0
P21	20	35	5	57.14	57.14	57.14	57.98	[31]	1.45
P22	24	40	7	100.0	100.0	100.0	100.0	[31]	0
P23	24	40	7	85.11	85.11	85.11	85.11	[31]	0
P24	24	40	7	73.51	73.51	73.51	73.51	[31]	0
P25	24	40	11	53.29	53.29	53.29	53.29	[31]	0
P26	24	40	12	48.61	48.95	48.88	48.95	[31]	0
P27	24	40	12	46.58	46.58	46.58	47.26	[31]	1.44
P28	27	27	5	54.82	54.82	54.82	54.82	[31]	0
P29	28	46	10	47.08	47.08	47.08	47.08	[18]	0
P30	30	41	14	62.94	63.31	63.27	63.31	[18]	0
P31	30	50	13	59.77	59.77	59.77	60.12	[20]	0.58
P32	30	50	14	50.56	50.83	50.80	50.83	[16]	0
P33	36	90	17	47.97	47.98	47.96	47.75	[8]	−0.48
P34	37	53	3	60.64	60.64	60.64	60.64	[31]	0
P35	40	100	10	84.03	84.03	84.03	84.03	[12]	0

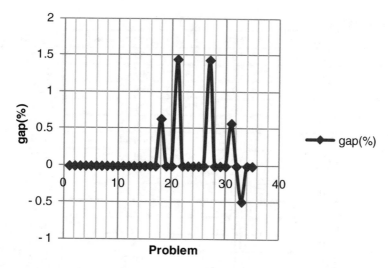

Fig. 3 The gap's variations

Indeed, the best and worst values are not equal for those problems. A statistical study is carried out thereafter to study the reliability of the average grouping efficiency. Jargon related to statistical fields should be introduced for a better comprehension. The word population is defined as the set of grouping efficiency obtained by applying the DGSA to randomly chosen individuals. The expression sample mean is defined as the average grouping efficiency of 10 runs. The mean of the overall population is defined as the average grouping efficiency of the individuals of search space after applying the DSGA.

The standard deviation, denoted SD, explains the distribution shape and the proximity of the grouping efficiency of individuals to the mean value. The SE represents the standard error in statistics. SE measure indicates how far the mean of the sample is from the mean of the whole population.

A smaller SE indicates that the sample mean is closer to the mean of the overall population. Table 3 shows that SE is equal to zero for twenty-nine instances. A SE equal to zero signifies that the sample mean is equal to the mean of the overall population. For the five remaining instances, the mean of the population is roughly equal to the sample mean minus or plus twice SE at 95 % confidence. Table 3 also contains the interval of the population mean for each instance and the SD for each problem.

The standard errors obtained in Table 3 are between 0 and 0.28. These errors are small and prove that the sample mean is not far from the overall population mean.

Table 3 Standard error

Problem	SD	SE	Lower bound	Upper bound
P1	0	0	82.35	82.35
P2	0	0	69.57	69.57
P3	0	0	79.59	79.59
P4	0	0	76.92	76.92
P5	0	0	60.87	60.87
P7	0	0	69.44	69.44
P9	0	0	58.72	58.72
P10	0	0	75.00	75.00
P11	0	0	92.00	92.00
P12	0	0	72.06	72.06
P13	0	0	71.83	71.83
P14	0	0	53.26	53.26
P15	0	0	69.53	69.53
P16	0	0	57.53	57.53
P17	0	0	57.73	57.73
P18	0.05	0.10	43.01	43.22
P19	0	0	50.81	50.81
P20	0	0	77.91	77.91
P21	0	0	57.14	57.14
P22	0	0	100.0	100.0
P23	0	0	85.11	85.11
P24	0	0	73.51	73.51
P25	0	0	53.29	53.29
P26	0.14	0.28	48.6	48.16
P27	0	0	46.58	46.58
P28	0	0	54.82	54.82
P29	0	0	47.08	47.08
P30	0.12	0.13	63.05	63.05
P31	0	0	59.77	59.77
P32	0.09	0.17	50.64	50.97
P33	0.03	0.06	47.9	48.02
P34	0	0	60.64	60.64
P35	0	0	84.03	84.03

7 Conclusion

In this chapter, the Cell Formation Problem and the mathematical model were introduced, then a brief presentation of the Discrete Gravitational Search Algorithm was performed, and, finally, the computational results obtained after the application of the discrete gravitational search algorithm to the cell formation problem were shown in the computational results section. In this work, the discrete gravitational search algorithm was applied for the first time to the cell formation problem with quite satisfactory results. A great number of modifications were necessary for successful application of the discrete gravitational search algorithm to the cell formation problem. Thus, the mutation operator of the dependent movement search phase was replaced by a crossover and the independent movement search encloses two mutation operators for more effective search space exploration. These changes improved the trade-off between exploration and exploitation. In addition, the changes reduced the computational time. The earliest gravitational search algorithm consumes a considerable computational time and suffers from a weak balance between exploration and exploitation. The enhanced discrete gravitational search algorithm proposed in this work goes beyond the limitations of the earliest one.

References

1. Akay, B., Karaboga, D.: Artificial bee colony algorithm for large-scale problems and engineering design optimization. J. Intell. Manuf. **23**(4), 1001–1014 (2010)
2. Burbidge, J.L.: The introduction of group technology. John Wiley & Sons, Incorporated, London (1975)
3. Chakraborty, P., et al.: On convergence of the multi-objective particle swarm optimizers. Inf. Sci. **181**(8), 1411–1425 (2011)
4. Dagli, C., Huggahalli, R.: Machine-part family formation with the adaptive resonance theory paradigm. Int. J. Prod. Res. **33**(4), 893–913 (1995)
5. Dimopoulos, C., Zalzala, A.M.S.: Recent developments in evolutionary computation for manufacturing optimization: problems, solutions, and comparisons. IEEE Trans. Evol. Comput. **4**(2), 93–113 (2000)
6. Doraghinejad, M., et al.: Channel assignment in multi-radio wireless mesh networks using an improved gravitational search algorithm. J. Netw. Comput. Appl. **38**, 163–171 (2014)
7. Dowlatshahi, M.B., et al.: A discrete gravitational search algorithm for solving combinatorial optimization problems. Inf. Sci. **258**, 94–107 (2014)
8. Elbenani, B., et al.: Genetic algorithm and large neighbourhood search to solve the cell formation problem. Expert Syst. Appl. **39**(3), 2408–2414 (2012)
9. Eusuff, M.M., Lansey, K.E.: Optimization of water distribution network design using the shuffled frog leaping algorithm. J. Water Resour. Plan. Manage. **129**(3), 10–25 (2003)
10. Farmer, J.D., et al.: The immune system, adaptation, and machine learning. Phys. D **2**(1–3), 187–204 (1986)
11. Goldengorin, B., et al.: The problem of cell formation: ideas and their applications. In: Cell Formation in Industrial Engineering. pp. 1–23. Springer, New York (2013)
12. Gonçalves, J.F., Resende, M.G.C.: An evolutionary algorithm for manufacturing cell formation. Comput. Ind. Eng. **47**(2–3), 247–273 (2004)

13. Gravel, M., Nsakanda, A.L.: Efficient solutions to the cell-formation problem with multiple routings via a double-loop genetic algorithm. Eur. J. Oper. Res. **109**(2), 286–298 (1998)
14. Holland, J.H.: Outline for a logical theory of adaptive systems. J. ACM **9**(3), 297–314 (1962)
15. Holland, J.H.: Adaptation in natural and artificial systems. University of Michigan Press, Ann Arbor, MI, USA (1975)
16. James, T.L., et al.: A hybrid grouping genetic algorithm for the cell formation problem. Comput. Oper. Res. **34**(7), 2059–2079 (2007)
17. Lei, D., Wu, Z.: Tabu search for multiple-criteria manufacturing cell design. Int. J. Adv. Manuf. Technol. **28**(9–10), 950–956 (2006)
18. Li, X., et al.: An ant colony optimization metaheuristic for machine-part cell formation problems. Comput. Oper. Res. **37**(12), 2071–2081 (2010)
19. Luo, J., Tang, L.: A hybrid approach of ordinal optimization and iterated local search for manufacturing cell formation. Int. J. Adv. Manuf. Technol. **40**(3–4), 362–372 (2008)
20. Mahdavi, I., et al.: Genetic algorithm approach for solving a cell formation problem in cellular manufacturing. Expert Syst. Appl. **36**(3), 6598–6604 (2009)
21. Mukattash, A.M., et al.: Heuristic approaches for part assignment in cell formation. Comput. Ind. Eng. **42**(2–4), 329–341 (2002)
22. Papaioannou, G., Wilson, J.M.: The evolution of cell formation problem methodologies based on recent studies (1997–2008): review and directions for future research. Eur. J. Oper. Res. **206**(3), 509–521 (2010)
23. Passino, K.M.: Biomimicry of bacterial foraging for distributed optimization and control. IEEE Control Syst. **22**(3), 52–67 (2002)
24. Rashedi, E., et al.: GSA: a gravitational search algorithm. Inf. Sci. **179**(13), 2232–2248 (2009)
25. Rezazadeh, H., et al.: Solving a dynamic virtual cell formation problem by linear programming embedded particle swarm optimization algorithm. Appl. Soft Comput. **11**(3), 3160–3169 (2011)
26. Shi, W., et al.: QSAR analysis of tyrosine kinase inhibitor using modified ant colony optimization and multiple linear regression. Eur. J. Med. Chem. **42**(1), 81–86 (2007)
27. Simon, D.: Biogeography-based optimization. IEEE Trans. Evol. Comput. **12**(6), 702–713 (2008)
28. Soleymanpour, M., et al.: A transiently chaotic neural network approach to the design of cellular manufacturing. Int. J. Prod. Res. **40**(10), 2225–2244 (2002)
29. Souilah, A.: Simulated annealing for manufacturing systems layout design. Eur. J. Oper. Res. **82**(3), 592–614 (1995)
30. Tian, H., et al.: Multi-objective optimization of short-term hydrothermal scheduling using non-dominated sorting gravitational search algorithm with chaotic mutation. Energy Convers. Manage. **81**, 504–519 (2014)
31. Tunnukij, T., Hicks, C.: An enhanced grouping genetic algorithm for solving the cell formation problem. Int. J. Prod. Res. **47**(7), 1989–2007 (2009)
32. Venkumar, P., Haq, A.N.: Complete and fractional cell formation using Kohonen self-organizing map networks in a cellular manufacturing system. Int. J. Prod. Res. **44**(20), 4257–4271 (2006)
33. Yang, X.-S.: Harmony search as a metaheuristic algorithm. In: Geem, Z.W. (ed.) Music-Inspired Harmony Search Algorithm, pp. 1–14. Springer, Berlin (2009)
34. Yang, X.-S., Deb, S.: Cuckoo search via Lévy flights. In: World Congress on Nature & Biologically Inspired Computing, NaBIC 2009, Coimbatore, India. pp. 210–214, 9–11 Dec 2009
35. Zolfaghari, S.: An objective-guided ortho-synapse Hopfield network approach to machine grouping problems

Parameterless Bat Algorithm and Its Performance Study

Iztok Fister Jr., Uroš Mlakar, Xin-She Yang and Iztok Fister

Abstract A parameter-free or parameterless bat algorithm is a new variant of the bat algorithm which was recently introduced. Characteristic of this algorithm is that user does not need to specify the control parameters when running this algorithm. Thus, this bat algorithm variant can have wide usability in solving real-world optimization problems. In this chapter, a preliminary study of the proposed parameterless bat algorithm is presented.

Keywords Bat algorithm · Control parameters · Optimization · Swarm intelligence

1 Introduction

As the global market becomes more competitive, it necessitates that companies should design their products and services in a more cost-effective and sustainable way so as to maximize their profits and performance and to minimize the costs and errors. All these must also meet the highest standards subject to various design constraints. This requires researchers and designers to follow the latest trends, to use the most cost-effective technologies and to use innovative optimization techniques. In recent years, artificial intelligence (AI) has started to become one of the most useful tools in industries and manufacturing engineering in the sense that artificial intelligence can help to design, develop, manufacture products more intelligently,

I. Fister Jr. (✉) · U. Mlakar · I. Fister
Faculty of Electrical Engineering and Computer Science,
University of Maribor, Smetanova 17, 2000 Maribor, Slovenia
e-mail: iztok.fister1@um.si

U. Mlakar
e-mail: uros.mlakar@um.si

I. Fister
e-mail: iztok.fister@um.si

X.-S. Yang
School of Science and Technology, Middlesex University, London NW4 4BT, UK
e-mail: x.yang@mdx.ac.uk

© Springer International Publishing Switzerland 2016
X.-S. Yang (ed.), *Nature-Inspired Computation in Engineering*,
Studies in Computational Intelligence 637, DOI 10.1007/978-3-319-30235-5_13

267

and such applications include robotics, car production, information technologies and obviously computer games.

Over the past few decades, many different optimization methods have been developed, and it is estimated there may be more than 100 different algorithms in the literature including the many variants of nature-inspired algorithms [7]. In fact, nature-inspired algorithms have become promising and powerful for solving the real-world optimization problems [14] and these algorithms often mimic the behavior of natural and biological systems.

In this chapter, we focus on the bat algorithm (BA) [13] and its extension to a parameter-free variant. In spite of its simplicity, this is a very efficient algorithm. The original BA has five parameters that represent a potential problem for users which usually do not know how to specify them properly. Therefore, this work introduces a new parameterless BA (i.e., PLBA) that eliminates this drawback by proposing techniques for a rational and automated parameter setting on behalf of the user. This study is an extension of our conference paper [6] where we presented a new variant of bat algorithm called parameterless BA (PLBA) that was based on the main idea of Lobo and Goldberg [11]. Here, the performance study has been extended in order to show the behavior of PLBA on dimensions higher than $D = 10$.

Therefore, the organization of this chapter is as follows. Section 2 provides the fundamentals of the bat algorithm. Section 3 discusses control parameters in the bat algorithm. Section 4 focuses on the experiments and presents the results obtained from experiments. Then the chapter concludes with the discussion of the performed work and some directions for the future are outlined.

2 Fundamentals of the Bat Algorithm

The standard bat algorithm was developed by Xin-She Yang in [13], and three idealized rules were used to capture some of the characteristics of microbats in nature:

- All bats use echolocation to sense the distance to target objects.
- Bats fly with the velocity v_i at position x_i, the frequency $Q_i \in [Q_{min}, Q_{max}]$ (also the wavelength λ_i), the rate of pulse emission $r_i \in [0, 1]$, and the loudness $A_i \in [A_0, A_{min}]$. The frequency (and wavelength) can be adjusted depending on the proximities of their targets.
- The loudness varies from a large (positive) A_0 to a minimum constant value A_{min}.

These idealized rules can be outlined as the pseudocode presented in Algorithm 1. The BA is a population-based algorithm with five parameters [6], and its population size is Np. The main loop of the algorithm (lines 4–16 in Algorithm 1) starts after the initialization (line 1), the evaluation of the generated solutions (line 2) and determination of the best solutions (line 4).

Algorithm 1 Bat algorithm

Input: Bat population $\mathbf{x_i} = (x_{i1}, \ldots, x_{iD})^T$ for $i = 1 \ldots Np, MAX\_FE$.
Output: The best solution $\mathbf{x}_{best}$ and its corresponding value $f_{min} = \min(f(\mathbf{x}))$.
1: init_bat();
2: $eval$ = evaluate_the_new_population;
3: f_{min} = find_the_best_solution($\mathbf{x}_{best}$); {initialization}
4: **while** termination_condition_not_met **do**
5:    **for** $i = 1$ **to** Np **do**
6:       $\mathbf{y}$ = generate_new_solution($\mathbf{x}_i$);
7:       **if** rand(0, 1) $< r_i$ **then**
8:          $\mathbf{y}$ = improve_the_best_solution($\mathbf{x}_{best}$)
9:       **end if** { local search step }
10:      f_{new} = evaluate_the_new_solution($\mathbf{y}$);
11:      $eval = eval + 1$;
12:      **if** $f_{new} \le f_i$ **and** $N(0, 1) < A_i$ **then**
13:         $\mathbf{x}_i = \mathbf{y}$; $f_i = f_{new}$;
14:      **end if** { save the best solution conditionally }
15:      f_{min}=find_the_best_solution($\mathbf{x}_{best}$);
16:    **end for**
17: **end while**

Briefly speaking, the BA algorithm consists of the following elements:

- generating a new solution (line 6),
- improving the best solution (lines 7–9),
- evaluating the new solution (line 10),
- saving the best solution conditionally (lines 12–14),
- determining the best solution (line 15).

The generation of a new solution obeys the following equations:

$$Q_i^{(t)} = Q_{min} + (Q_{max} - Q_{min})N(0, 1),$$
$$\mathbf{v}_i^{(t+1)} = \mathbf{v}_i^t + (\mathbf{x}_i^t - \mathbf{x}_{best})Q_i^{(t)}, \tag{1}$$
$$\mathbf{x}_i^{(t+1)} = \mathbf{x}_i^{(t)} + \mathbf{v}_i^{(t+1)},$$

where $N(0, 1)$ is a random number drawn from a uniform distribution, and $Q_i^{(t)} \in [Q_{min}^{(t)}, Q_{max}^{(t)}]$ is the frequency determining the magnitude of the velocity change. The improvement of the current best solution is performed according to the following equation:

$$\mathbf{x}^{(t)} = \mathbf{x}_{best} + \epsilon A_i^{(t)} N(0, 1), \tag{2}$$

where $N(0, 1)$ denotes the random number drawn from a Gaussian distribution with a zero mean and a standard deviation of one. In addition, ϵ is the scaling factor and $A_i^{(t)}$ the loudness. The improvement of the current best solution is controlled by a parameter r_i.

It is worth pointing out that these parameters can somehow balance the exploration and exploitation components of the BA search process, where exploitation is governed by Eq. (1) and exploitation by Eq. (2).

The evaluation function models the characteristics of the problem to be solved. The archiving of the best solution conditionally is similar to that used in simulated annealing (SA) [10], where the best solution is taken into the new generation according to a probability controlled by the parameter A_i. In fact, the parameter A_i prevents the algorithm to get stuck into a local optimum.

The bat algorithm has attracted a lot of interests in the literature and has been applied to many applications. For a relatively comprehensive review, please refer to Yang and He [15].

2.1 How to Control Parameters in Bat Algorithm?

As is the case for all evolutionary algorithms (EA) [4] and swarm-intelligence based (SI-based) algorithms [1], they are stochastic algorithms and all these algorithms have algorithm-dependent parameters. The values of these parameters will largely affect the performance of the algorithm under consideration. To control these parameters is equivalent to guiding the algorithm's search process during the exploration of the search space of an optimization algorithm and hence may have a huge influence on the quality of obtained solutions. Thus, parameter tuning and control become an important topic for active research, though the setting of these parameters may depend on the type of problems and may be well problem-specific. For example, the bat algorithm is guided by five parameters, and the following parameters are the part of basic bat algorithm:

- the population size NP,
- loudness A_i,
- pulse rate r_i,
- minimum frequency Q_{min} and
- maximum frequency Q_{max}.

In order to avoid the tedious task for tuning parameters, a parameterless variant of the bat algorithm has been proposed here and will be presented in detail in the rest of this chapter.

3 Design of a Parameterless Bat Algorithm

In order to develop a new parameterless BA (PLBA), the influence of these algorithm-dependent parameters was studied and extensive studies revealed that some algorithm parameters can be rationally set, while the optimal setting of another parameters needs to be found automatically. For example, the parameter $Q_i \in [Q_{min}, Q_{max}]$

determines the magnitude of the change and settings of parameters Q_{min} and Q_{max} depend on the problem of interest. However, the rational setting of this parameter can be approximated with the lower x_i^{Lb} and upper x_i^{Ub} bounds of the particular decision variables as follows:

$$Q_i^{(t)} = \frac{x_i^{(Ub)} - x_i^{(Lb)}}{Np} \cdot N(0, 1), \tag{3}$$

where $N(0, 1)$ has the same meaning as in Eq. (1). For example, when the $x_i^{(Lb)} = -100.0$ and $x_i^{(Ub)} = 100.0$, the frequency is obtained in the interval $Q_i^{(t)} \in [0, 2]$.

The rationality for setting the values of parameters r_i and A_i is based on the following consideration. Parameter r_i controls the exploration/exploitation components of the BA search process. The higher the value, the more the process is focused on the exploitation. However, the higher r_i also means that the modified copies of the best solution are multiplied in the neighborhood of the current population. As a result of high r_i, premature convergence to the local optimum may occur. Therefore, the rational setting of this parameter is to set r_i properly, and extensive numerical experiments suggested $r_i = 0.1$ is a good value that improves on average one in ten solutions in the current population.

Similar consideration is valid also for parameter A_i. The lower the parameter, the less time the best current solution is preserved in the new generation. Therefore, the rational selection of this parameter is $A_i = 0.9$ that preserves the 90% of the best solutions and ignores the remaining 10%.

The population size is also a crucial parameter in the BA. The lower population size may suffer from the lack of diversity, while a higher population size may cause slow convergence. However, the rational setting of the population size depends on the problem of interest. Therefore, an automatic setting of this parameter is proposed in the PLBA, where the population size is varied in the interval $Np \in [10, 1280]$ such that each population size is multiplied by two in each run starting with $Np = 10$. As a result, eight instances of the PLBA denoted as PL-1 to PL-8 are obtained and a user needs to select the best results among the instances generated.

4   Experiments and Results

The purpose of our experimental work was to show that the results of the proposed PLBA are comparable if not better than the results of the original BA. In line with this, the original BA was compared with eight instances of the PLBA using the rational and automatic parameter setting; i.e., Q_i was calculated according to Eq. (3) (e.q., $Q_i \in [0.0, 2.0]$), pulse rate and loudness were fixed as $r_i = 0.1$ and $A_i = 0.9$, while the population size was varied in the interval $Np = \{10, 20, 40, 80, 160, 320, 640, 1280\}$. In contrast, the original BA was run using the following parameters: $Np = 100$, $r_i = 0.5$, $A_i = 0.5$, and $Q_i \in [0.0, 2.0]$.

Table 1 Summary of the benchmark functions

Tag	Function name	Definition
f_1	Ackley's	$-20\exp(-0.2\sqrt{\frac{1}{n}\sum_{i=1}^n x_i^2}) - \exp(\frac{1}{n}\sum_{i=1}^n \cos(2\pi x_i)) + 20 + e$
f_2	Griewank	$1 + \frac{1}{4000}\sum_{i=1}^n x_i^2 - \prod_{i=1}^n \cos(\frac{x_i}{\sqrt{i}})$
f_3	Rastrigin	$10n + \sum_{i=1}^n (x_i^2 - 10\cos(2\pi x_i))$
f_4	Sphere	$\sum_{i=1}^n x_i^2$
f_5	Whitley	$\sum_{i=1}^n \sum_{j=1}^n \left[\frac{(100(x_i^2-x_j)^2+(1-x_j)^2)^2}{4000} \right.$ $\left. - \cos(100(x_i^2 - x_j)^2 + (1 - x_j)^2) + 1 \right]$
f_6	Exponential_function	$\sum_{i=1}^n e^{ix_i} - \sum_{i=1}^n e^{-5.12i}$
f_7	Quartic	$\sum_{i=0}^n ix_i^4 + \text{random}[0,1)$
f_8	Ridge	$\sum_{i=1}^n (\sum_{j=1}^i x_j)^2$
f_9	Schwefel	$\sum_{i=1}^n (-x_i \sin(\sqrt{|x_i|})) + 418.982887 \cdot n$
f_{10}	Double_Sum	$\sum_{i=1}^n (\sum_{j=1}^i (x_j - j)^2)$

Table 2 Function domain

Function	Domain
f_1	$[-32, 32]$
f_2	$[-512, 512]$
f_3	$[-5.12, 5.12]$
f_4	$[-100, 100]$
f_5	$[-10.24, 10.24]$
f_6	$[-5.12, 5.12]$
f_7	$[-1.28, 1.28]$
f_8	$[-64, 64]$
f_9	$[-512, 512]$
f_{10}	$[-10.24, 10.24]$

Interestingly, the setting of frequency in the original BA matches the rational setting in PLBA according Eq. (3).

Experiments were run on the benchmark suite consisting of ten functions presented in Table 1.

Test functions and the domains of parameters shown in Table 2 were mainly based on the Neal Holtschulte website [9].

The dimensions of functions were set to $D = 10, D = 20, D = 30$. All algorithms stopped after $FE = 1000 * D$ of fitness function evaluations and each algorithm was run 25 times. Three tests have been conducted according to each of the dimension used. It is worth pointing out that there are different suggestions concerning the number of function evaluations in the literature and most studies use far more numbers of evaluations such as $100,000D$. However, as our purpose here is to mainly show this parameter-free bat algorithm works, we will use far fewer numbers of evaluations. Obviously, the results will in general improve as the number of function evaluations increases.

In the remainder of the chapter, the results of the benchmark functions are presented in detail. In addition, statistical tests are illustrated in order to show the differences and to see how each algorithm performs. The results for the case of $D = 10$ suggested that the results are improved by increasing the population size. The best results were obtained by the biggest population size (i.e., $D = 1280$).

In order to estimate the results statistically, Friedman tests [8] were conducted. In essence, this Friedman test is a two-way analysis of variances where the test statistic is calculated and then converted to ranks, followed by post-hoc tests using the calculated ranks. Loosely speaking, a low value of rank means a better algorithm [3]. The post-hoc tests are performed only if a null hypothesis of the Friedman test is rejected. In short, the so-called null hypothesis corresponds to the case that medians between ranks of all algorithms are equal. According to Demšar [2], the Friedman test is a safe and robust non-parametric test for comparing more algorithms over multiple data sets, and this test, together with the corresponding Nemenyi post-hoc test, can ensure a neat presentation of statistical results [12]. This test have been conducted using a significance level 0.05 in this study.

Overall, statistical tests can typically reflect the performance of the algorithms against the benchmarks. In summary, nine classifiers (i.e., the results according to different population sizes) were compared, leading to $10 * 5 = 50$ variables, where the first number in the expression denotes the number of functions and the second the number of measures (i.e., minimum, maximum, mean, median and standard deviation). Three Friedman tests were performed regarding the different dimensions.

The results of the Friedman non-parametric test for dimension $D = 10$ are presented in Fig. 1 with a table and a diagram dedicated to the dimension under consideration. In the table, the results of Friedman tests are given together with corresponding Nemenyi post-hoc test. The results of Nemenyi post-hoc test are presented as intervals of critical differences. The best algorithm in Nemenyi post-hoc test is denoted with

Alg.	Friedman	Nemenyi	
		Critical difference	Sign
PL-1	8.41	[7.57,9.25]	†
PL-2	7.22	[6.38,8.06]	†
PL-3	6.56	[5.72,7.40]	†
PL-4	5.37	[4.53,6.21]	†
PL-5	4.12	[3.28,4.96]	†
PL-6	3.34	[2.50,4.18]	†
PL-7	2.25	[1.41,3.09]	
PL-8	1.43	[0.59,2.27]	‡
BA	4.86	[4.02,5.70]	†

(a) $D = 10$

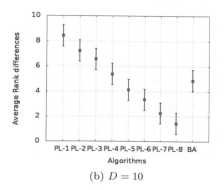

(b) $D = 10$

Fig. 1 Statistical analysis of the results by optimizing functions $D = 10$

‡ symbol in the table, while the significant difference between the control method and corresponding algorithm is depicted by † symbol.

Schematically, the results of the Nemenyi post-hoc test are illustrated in a corresponding diagram. In the diagram, the points show the average ranks and the lines indicate the confidence intervals (critical differences) for the algorithms under consideration. The lower the rank value, the better the algorithm. On the other hand, two algorithms are said to be significantly different when their intervals do not overlap.

From Fig. 1, it can be seen that the PLBA with $Np = 1280$ (i.e., PL-8) is really the best instance according to the Friedman test. This instance outperformed all the other instances of the original BA and the PLBAs except for the case with a population size of $Np = 640$ (i.e., PL-7).

However, it is worth pointing out that the standard BA performs almost equally well, compared to the PLBA-4 with the population size of $Np = 80$. With the actual population size of $Np = 100$ in BA, this means that a moderate size can be sufficient to get very good results, not the necessarily the best results. This result is consistent with our observations and other research in the literature, which suggests that the results usually improve the population size increases for most population-based algorithms. However, there are some algorithms that may not be very sensitive to the population size. For example, the firefly algorithm usually can obtain good results for small and moderate population size $NP = 20$ to $NP = 100$ [5]. Therefore, the effect of the population size needs further investigation for nature-inspired algorithms.

Furthermore, the optimization results of the benchmark functions for $D = 20$ and $D = 30$ show similar trends to the case of dimension $D = 10$. For example, for $D = 30$, the Friedman tests are summarized in Fig. 2 where similar trends are observed.

In summary, the experiments showed that the original BA algorithm can obtain good results and its performance usually increase as the population size increases. On the other hand, the convergence rate of this algorithm can be very high and thus it may lead to premature convergence if the population size is too low as the algorithm can quickly converge to local optima. It seems that the parameterless BA algorithm

Alg.	Friedman	Nemenyi	
		Critical difference	Sign
PL-1	8.11	[7.27,8.95]	†
PL-2	7.38	[6.54,8.22]	†
PL-3	6.62	[5.78,7.46]	†
PL-4	5.34	[4.50,6.18]	†
PL-5	3.78	[2.94,4.62]	†
PL-6	2.94	[2.10,3.78]	
PL-7	2.13	[1.29,2.97]	
PL-8	1.62	[0.78,2.46]	‡
BA	4.65	[3.81,5.49]	†

(a) $D = 30$

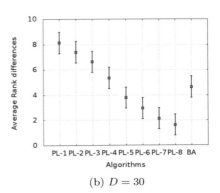

(b) $D = 30$

Fig. 2 Statistical analysis of the results by optimizing functions $D = 30$

can avoid this problem using the higher population sizes. However, selecting the best population size during this automatic search for the proper parameter setting may still depend on the type of problems under consideration.

In addition, though the settings are fixed for A_i and r_i, it seems that a better approach for parameter-free variants should be adaptive; that is to vary A_i and r_i in an adaptive manner so that users need not to worry about the setting of the algorithms, and this adaptivity should also include the population size. More studies are needed to achieve these goals.

5 Conclusion

This chapter presented a performance study of recently proposed parameterless bat algorithm. This variant of the bat algorithm does not need to set the control parameters. Experiments were conducted on a set of ten standard benchmark functions with three different dimensions. In a nutshell, experiments confirmed that parameterless bat algorithm is a very promising variant with the potential to solve the real-world optimization applications.

However, this preliminary study may also suggest that other population-based algorithms can have similar effects of population sizes and the existing literature about particle swarm optimization seemed also to suggest that results can improve significantly as the population size increases. That is partly the reason that it is suggested the population size should be increased for high-dimensional problems.

It is worth pointing out that this parameterless variant of the bat algorithm used fixed parameters A_i and r_i, while allowing the variation of the population size. It can be expected that a more robust approach is to use adaptive settings to set all the parameters including the population size automatically for a given set of problems. But how to achieve this is still an open questions. There are some other open problems regarding automatic parameter tuning and control. Therefore, more detailed studies are highly needed to systematically analyze and study all major population-based metaheuristic algorithms.

References

1. Blum, C., Li, X.: Swarm intelligence in optimization. In: Blum, C., Merkle, D. (eds.) Swarm Intelligence: Introduction and Applications, pp. 43–86. Springer, Berlin (2008)
2. Demšar, J.: Statistical comparisons of classifiers over multiple data sets. J. Mach. Learn. Res. **7**, 1–30 (2006)
3. Derrac, J., Garca, S., Molina, D., Herrera, F.: A practical tutorial on the use of nonparametric statistical tests as a methodology for comparing evolutionary and swarm intelligence algorithms. Swarm Evol. Comput. **1**(1), 3–18 (2011)
4. Eiben, A.E., Smith, J.E.: Introduction to Evolutionary Computing. Springer, Berlin (2003)
5. Fister, I., Yang, X.-S., Brest, J., Fister Jr., I.: Modified firefly algorithm using quaternion representation. Expert Syst. Appl. **40**(18), 7220–7230 (2013)

6. Fister Jr., I., Fister, I., Yang, X.-S.: Towards the development of a parameter-free bat algorithm. In: StuCoSReC: Proceedings of the 2015 2nd Student Computer Science Research Conference, pp. 31–34 (2015)
7. Fister Jr., I., Yang, X.-S., Fister, I., Brest, J., Fister, D.: A brief review of nature-inspired algorithms for optimization. Elektrotehniški vestnik **80**(3), 116–122 (2013)
8. Friedman, M.: A comparison of alternative tests of significance for the problem of m rankings. Ann. Math. Stat. **11**, 86–92 (1940)
9. Holtschulte, N., Moses, M.: Should every man be an island. In: GECCO 2013 Proceedings, 8 pp. (2013)
10. Kirkpatrick, S., Gelatt, C.D., Vecchi, M.P.: Optimization by simulated annealing. Science **220**(4598), 671–680 (1983)
11. Lobo, F.G., Goldberg, D.E.: An overview of the parameter-less genetic algorithm. In: Proceedings of the 7th Joint Conference on Information Sciences (Invited paper), pp. 20–23 (2003)
12. Nemenyi, P.B.: Distribution-free multiple comparisons. Ph.D. thesis, Princeton University (1963)
13. Yang, X.-S.: A new metaheuristic bat-inspired algorithm. In: Nature Inspired Cooperative Strategies for Optimization (NICSO 2010), pp. 65–74. Springer (2010)
14. Yang, X.-S.: Nature-Inspired Optimization Algorithms. Elsevier (2014)
15. Yang, X.-S., He, X.: Bat algorithm: literature review and applications. Int. J. Bio-inspir. Comput. **5**(3), 141–149 (2013)

Printed in the United States
By Bookmasters